LIQUIDS AND THEIR PROPERTIES

A Molecular and Macroscopic Treatise
with Applications

MATHEMATICS & ITS APPLICATIONS

Series Editor: Professor G. M. Bell

Chelsea College, University of London

Mathematics and its applications are now awe-inspiring in their scope, variety and depth. Not only is there rapid growth in pure mathematics and its applications to the traditional fields of the physical sciences, engineering and statistics, but new fields of application are emerging in biology, ecology and social organisation. The user of mathematics must assimilate subtle new techniques and also learn to handle the great power of the computer efficiently and economically.

The need of clear, concise and authoritative texts is thus greater than ever and our series will endeavour to supply this need. It aims to be comprehensive and yet flexible. Works surveying recent research will introduce new areas and up-to-date mathematical methods. Undergraduate texts on established topics will stimulate student interest by including applications relevant at the present day. The series will also include selected volumes of lecture notes which will enable certain important topics to be presented earlier than would otherwise be possible.

In all these ways it is hoped to render a valuable service to those who learn, teach, develop and use mathematics.

LIQUIDS AND THEIR PROPERTIES

A MOLECULAR AND MACROSCOPIC TREATISE WITH APPLICATIONS

H. N. V. Temperley, Sc.D. (Cantab)
Professor of Applied Mathematics, University College, Swansea
formerly Fellow of King's College, Cambridge

and

D. H. Trevena, Ph.D. (Wales), Ph.D. (Cantab)
Senior Lecturer in Physics, University College of Wales, Aberystwyth
formerly Fellow of the University of Wales

ELLIS HORWOOD LIMITED
Publisher   Chichester

Halsted Press: a division of
JOHN WILEY & SONS
Chichester · New York · Brisbane · Toronto

The publisher's colophon is reproduced from James Gillison's ancient Market Cross, Chichester

First published in 1978 by

ELLIS HORWOOD LIMITED

Market Cross House, 1 Cooper Street, Chichester, Sussex, England

Distributors:

Australia, New Zealand, South-east Asia:
Jacaranda-Wiley Ltd., Jacaranda Press,
JOHN WILEY & SONS INC.,
G.P.O. Box 859, Brisbane, Queensland 4001, Australia.

Canada:
JOHN WILEY & SONS CANADA LIMITED
22 Worcester Road, Rexdale, Ontario, Canada.

Europe, Africa:
JOHN WILEY & SONS LIMITED
Baffins Lane, Chichester, Sussex, England.

North and South America and the rest of the world:
HALSTED PRESS, a division of
JOHN WILEY & SONS
605 Third Avenue, New York, N.Y. 10016, U.S.A.

© 1978 H. N. V. Temperley and D. H. Trevena/Ellis Horwood Limited

Library of Congress Cataloging in Publication Data
Temperley, H. N. V.
    Liquids and their properties.
    (Mathematics and its applications)
    1. Liquids. I. Trevena, D. H., joint author.
II. Title.
QC145.2.T45   530.4'2   77-8177

ISBN 85312-065 (Ellis Horwood Ltd.)
ISBN 0-470-99203-4 (Halsted Press)

Typeset in Press Roman by Coll House Press, Chichester, Sussex

Printed by Cox and Wyman Limited, Fakenham.

DEDICATION

We dedicate this book to the memory of Professor John Gamble Kirkwood of Yale University. He founded the theory of the liquid distribution function in 1935, and the microscopic theory of irreversible processes in 1946, and inspired many subsequent workers.

Table of Contents

Authors' Preface

Liquids and their properties was written to fill a gap affecting several disciplines and whose existence becomes increasingly obvious in a number of fields. Our fundamental knowledge of the properties of liquids and solutions has advanced considerably in the years since 1957 when it was conclusively demonstrated that modern computers are capable of handling the formidable mathematical problems associated with the statistical mechanics of liquids. The theory of the liquid distribution function had been previously created by Professor J. G. Kirkwood and his school in a series of papers beginning in 1935.

The book is designed to be of use to students of both the pure and applied sciences at all levels and includes material of a standard suitable for the appropriate university honours courses, and also specialised courses in polytechnics. Research workers, in teaching or industry, who require a basic knowledge of the liquid phase may use it as a reference book. A novel and important feature of this book lies in the emphasis on both the molecular and macroscopic approaches to liquids and the demonstration of how they complement each other. The book has evolved from university lecture courses which we have both given over a number of years, from our research work on various aspects of the liquid phase, and from our contacts with industry.

Since the properties of liquids and solutions are of importance in all of the sciences and in most of the technologies, we believe that an account of the liquid phase has strong claims to importance in universities, polytechnics and medical schools for courses, not only in physics but also in applied mathematics, chemistry, metallurgy, engineering, biology and medicine. The recent advances in our fundamental knowledge of liquids have already had some impact on metallurgy, chemistry and chemical engineering, and will certainly spread to these other cognate subjects before long. We are only too well aware that worthy topics are often crowded out of syllabuses, and so we offer this book as what every good scientist, whether in research, teaching, industry or medicine should know about liquids. Consequently we have included chapters on hydrodynamics and ultrasonics, both because they satisfy this criterion and because they are often neglected in physics courses despite their increasing technical importance.

We are very conscious of the direct and indirect help that we have received, especially from our students and university colleagues. For various reasons, the writing has been spread over a long time and so our helpers are too numerous to mention by name, but we have made every effort to acknowledge published contributions to the field and apologise in advance for any errors or omissions here. We have tried to be as factual as possible about controversial topics, which exist in this field as in all other rapidly changing ones.

It is a great pleasure to thank our publisher Mr Ellis Horwood and his colleagues for the very friendly co-operation and patience which they have shown us throughout the publishing of this book. We would also like to take this opportunity to express our thanks to Professor G. M. Bell, Chelsea College, London University, the series editor of *Mathematics and its Applications,* for his valuable suggestions and help towards the planning and writing of this book from the manuscript stage through to its printing and publication.

H.N.V.T. would like to put on record his immense debt to Prof. J. G. Kirkwood of Yale University. Progress in the fundamental theory of liquids was slow until the advent of large computers and was also delayed by the war years and by the diversion of Kirkwood to work on underwater explosions in the U.S.A. H.N.V.T. took part in the U.K. side of this work, and, although the approaches were slightly different, they showed that the experimental facts about underwater explosions and damage to ships could be understood using only thermodynamics, hydrodynamics and the theory of structures. In certain circumstances, underwater explosions can give rise to tensions in water. A study of the effects of these led us to take up the study of the behaviour of liquids under tension (see Chapter 8). The phenomenon of cavitation is important in the design of, e.g., propellers, hydraulic equipment and lubricated bearings, and consequently has a very large literature. Studies of cavitation and of the fundamental theory of liquids have helped one another considerably. In 1952–3 H.N.V.T. had the privilege of working with Kirkwood at Yale and will always value his encouragement and support at an extremely critical time in his career. We dedicate this book to his memory.

H.N.V.T.
D.H.T.
August 1977

Glossary of Symbols and Physical Constants

A	Area	F	Force
A	Helmholtz free energy	F	Tension in a liquid
A_0	Wave amplitude	$\hat{F}$	Breaking tension of a liquid
a	Constant in van der Waals equation	$F(r)$	Intermolecular force between two molecules at separation r
a	Acceleration	f	Partition function for single particle in an assembly
a	Radius of bubble		
a	Radius of 'sphere of influence' around a molecule in a liquid	f	Frequency
		f_0	Relaxation frequency
		f_{ij}	Mayer f-function
B	Second virial coefficient	f_N	Complete distribution function for the coordinates and momenta of N molecules in a volume V
b	Constant in van der Waals equation		
b_ℓ	Mayer cluster integral for ℓ molecules		
		G	Gibbs free energy
C	Third virial coefficient	G^*	Complex shear modulus
C_A	Heat capacity of liquid at constant surface area	G'	Elastic modulus
		G''	Viscous modulus
C_D	Drag coefficient	$g(r)$	Radial distribution function
C_p, C_v	Specific heats at constant pressure and constant volume	h	Planck's constant
		I	Intensity of wave
c	Velocity of sound	$I(S)$	Scattered intensity of X-rays or neutrons
c_1, c_2	Velocities of first and second sound in liquid Helium II		
		K, K_S, K_T	Bulk modulus, adiabatic and isothermal bulk modulus
D	Fourth virial coefficient		
d	Packing density of assembly of balls	k	Boltzmann constant
		k	Wave number $(2\pi/\lambda)$
d	Spacing in diffraction grating	ℓ_f, ℓ_v	Latent heat of fusion, of vaporization
E	Electric field		
E	Total surface energy per unit area of liquid surface	M	Molecular weight
		M_A, M_B	Number of molecules of liquids A, B in a binary liquid
e	Electronic charge		

	mixture
m	Mass
N	Number of particles (molecules) in an assembly
N_A	Avogadro constant (number)
N_A, N_B	Number of molecules of types A, B in a mixed vapour phase
n_2	Pair distribution function
n_A, n_B	Number of moles of solvent and solute in AB solution
P	Pressure
P	Macroscopic momentum flow
P_c, P_r	Critical pressure, triple point pressure
P^*,etc.	Reduced pressure, etc.
P_A, P_B	Partial vapour pressures of components A, B in an AB liquid mixture
P_A^0, P_B^0	Vapour pressures of pure A, B components
P_{xy}	Shear stress in the x-direction perpendicular to the y-axis
P'	Parachor
p	Excess or acoustic pressure
p	Momentum (components p_x, p_y, p_z)
Q	Quantity of heat
Q	Configurational partition function
Q	Macroscopic energy flow
Q	Volume rate of flow of liquid through a tube
R	Universal gas constant
R	Radius of curvature of surface
r	Intermolecular separation
r	Radial distance
r_0	Normal equilibrium intermolecular separation
r_{ij}	Separation of two 'labelled' molecules i, j
r, θ, ϕ	Spherical polar coordinates

S	Entropy
S	Strain
S	Spreading coefficient
S	Rate of shear
$S(k, \omega)$	Intensity of scattered neutrons as a function of scattering angle and momentum
s	Condensation $((\rho - \rho_0)/\rho_0)$
T	Thermodynamic (absolute) temperature
T_c, T_r	Critical temperature, triple point temperature
T_f, T_b	Filling and breaking temperatures of a liquid in a Berthelot tube
t	time
U	Internal energy
U	Velocity
u	Particle velocity
V	Volume
V_c	Critical volume
v	Velocity
v	Volume
v_f	Free volume
W	Work
W	Number of complexions of an assembly of particles
W_A, W_B	Work of cohesion of pure liquids A and B
W_{AB}	Work of adhesion between two liquids A and B
W_{SL}	Work of adhesion between liquid L and solid S
x, y, z	Cartesian coordinates
x_A, x_B	Mole fractions of components A, B in an AB liquid mixture
Z	Partition function for an assembly of particles
z	Number of nearest neighbours (coordination number) in a solid (lattice) or a liquid (quasi-lattice)
$\not{z}$	Fugacity

α	Dielectric polarizability	ρ	Density
α	Absorption coefficient	ρ_ℓ, ρ_v	Density of liquid, density of vapour
α'	Absorption coefficient expressed in decibels per metre	ρ_n, ρ_s	Densities of normal and superfluid components of liquid Helium II
β	Thermal expansion coefficient (thermal expansivity)	$\rho(r), \rho_0$	Number density and mean number density of molecules
β	Yield stress	ρc	Acoustic impedance
β_k	Mayer irreducible cluster integral for $(k+1)$ molecules	σ	Molecular diameter
γ	Free surface energy, surface tension	τ	Time
		τ	Relaxation time
γ	Ratio of specific heats, C_p/C_v	τ	Shear stress
δ	Logarithmic decrement	Φ	Total potential energy of assembly of particles
δ	Phase angle difference		
ϵ	Energy	Φ	Lattice potential energy per mole
ϵ	Maximum (negative) value of intermolecular potential energy (dissociation energy)	ϕ	Velocity potential
		$\phi(r)$	Interaction energy between two molecules at separation r (pair potential function)
η	Viscosity		
η_a	Apparent viscosity		
η^*	Complex dynamic viscosity	ϕ_A, ϕ_R	Attractive and repulsive energies between two molecules
Θ	Debye characteristic temperature	ω	Pulsatance or angular frequency $(2\pi f)$
θ	Angle of contact		
κ	Thermal conductivity		
κ_T	Isothermal compressibility		
λ	De Broglie wavelength (h/mv)		
λ	Wavelength		
μ	Moment of electric dipole	e	Electronic charge 1.602×10^{-19} C
μ	Chemical potential		
μ	Bingham viscosity	h	Planck's constant 6.626×10^{-34} Js
ν	Frequency		
ν	Kinematic viscosity	k	Boltzmann's constant 1.381×10^{-23} JK^{-1}
ξ	Particle displacement		
ξ	Autocorrelation function of force $F(t)$	N_A	Avogadro constant (number) 6.022×10^{23} mole^{-1}
π	Surface pressure	R	Universal gas constant 8.314 J mole^{-1} K^{-1}
Π	Osmotic pressure		

Chapter 1
HISTORICAL INTRODUCTION

Interest in liquids is as old as science itself, and it has always been realised that their properties are of significance in almost all sciences. The ancients had the aeropile, a forerunner of the steam turbine, and invented various devices like the water-clock and the coin-in-the-slot machine that exploited the flow properties of liquids. They were aware of the medical and biological importance of water and solutions and of the parts played by water and ice in the erosion of mountains and in the deposition of silt. At a very early stage in their development men discovered, probably empirically, the properties of liquid metals and alloys. Modern engineering metallurgy is still very largely based on crafts such as casting, alloying and hot and cold working of metals.

The most striking difference between a liquid and a solid is that a liquid does not permanently resist forces tending to change its shape. This distinction is still good enough for most practical purposes though, as usual, doubtful intermediate cases can be found. Calculus and mechanics emerged during the Newtonian epoch and, even though the molecular structure of matter was then only a vague speculation, it was natural to try to work out the mechanics of a **continuous medium**. These efforts led to the creation of elasticity and hydrodynamics, with special reference to solids, which resist small shearing stresses permanently, and to liquids which do not. Much of the hydrodynamics originally laid down by Newton, Bernoulli, Euler and Lagrange remains perfectly valid today. The modifications required to allow for the fact that matter has a discontinuous molecular structure can be important for low-pressure gases, but are of little significance for liquids where the molecules are nearly in contact.

Matters remained in this state until the molecular theory of matter was developed in the early nineteenth century. This led on to the question of the relationships between the properties of individual molecules and the 'bulk' properties of matter (such as the heat capacity and the compressibility and other elastic moduli, ususally described as equilibrium properties) with others like thermal and electrical conductivity and diffusivity, usually described as transport properties because they can only be measured under non-equilibrium conditions. If we consider mixtures of two different types of molecule, the number of bulk properties becomes much greater. We are now interested in further equilibrium properties like solubility and in further transport properties like the rate of diffusion of one substance into another. We also meet a large number of cross-effects. If, for example, a mixture is subjected to a temperature-gradient and a concentration gradient together, the flows of heat and of matter are not independent of each other. A flow of heat through an originally uniform mixture can, in certain circumstances, give rise to a concentration gradient. This is a typical example of a cross-effect.

Quite early it was recognised that molecules must have nearly hard cores and also that they must attract one another. In no other way does it seem possible to account both for the cohesion of solids and for the fact that their densities change relatively slowly with temperature. Expressed in terms of the mutual potential energy of two molecules we must expect, for the idealised case of nearly spherical molecules (easiest to treat theoretically, and reasonably well reproduced in practice by rare-gas atoms like neon and argon), a relationship between potential energy and distance between centres of the general type shown in Fig. 1.1. A large energy is required to push the centres of two molecules much closer than OA, usually known as the 'gas-kinetic diameter'; but, at somewhat longer distances, the force becomes weakly attractive. For molecules that cannot be regarded as approximate spheres, we may need more than just the distance apart of their centres of mass to describe their relative geometrical

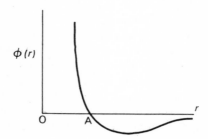

Fig. 1.1 Interaction energy of two molecules as a function of distance.

positions properly; for example, for diatomic molecules we need to know the relative directions of their axes. It remains universally true that the interaction energy is repulsive at very short distances and attractive at somewhat longer distances.

For an explanation of such interactions we must turn to quantum mechanics. It is possible to understand Fig. 1.1 qualitatively in terms of the quantum-mechanical picture of an atom or molecule. The molecule is made up of positively charged nuclei each surrounded by an electron-cloud. At large distances the mutual perturbations of the electron clouds of two molecules which do not have permanent dipole moments give rise to the so-called London attraction discussed in Chapter 3 while the corresponding energy is expected to vary as the inverse sixth power of the distance. (For a reasonably simple molecule, reliable calculations of this energy can be made in terms of its optical and infrared spectroscopic properties.) As the two molecules are pushed closer together the electrostatic repulsions between the nuclei begin to be important in addition to mutual distortion of the electron clouds around the nuclei. The net result is a repulsion, but this is much harder to calculate numerically than is the attraction and is sometimes represented as a power law. Purely theoretical calculations can only be attempted for relatively simple molecules. These matters are further discussed in Chapter 3.

The assumption of effectively spherical molecules does not restrict us to the rare gases which, according to quantum mechanics, have spherically symmetrical atoms. There are plenty of molecules, like carbon tetrachloride and benzene, which present a relatively uniform 'appearance' to the outside world. In liquids, it is very likely that such molecules are rotating, in which case directional effects can be largely ignored. As long as a liquid was thought of as having a continuous structure, one could do little more than accept its physical properties such as density and viscosity without trying to explain them. Once we introduce a molecular hypothesis we ask to what extent we can relate the molecular interaction function shown in Fig. 1.1 to the macroscopic properties of liquids. Similar questions can be asked for gases and solids, and we now trace briefly the history of the kinetic theories of gases, solids and liquids.

Once it was realised by Joule, Rumford and others that heat and molecular motion are identical, physicists met the problem of describing, and predicting, the macroscopic properties of a large assembly of molecules in ceaseless motion endlessly colliding with one another. As is shown in treatises on statistical mechanics, it is not necessary actually to solve the equations of motion of the individual molecules in order to derive equilibrium properties. The easiest assembly to study is a gas, and this was accordingly tackled first. Simplifications occur because, at ordinary densities, the spacing between neighbouring molecules is many times the gas-kinetic diameter. A typical molecule spends most of its time 'running free' between collisions and so triple and quadruple collisions are very rare.

The kinetic theory of gases seems to have been developed independently by Clerk-Maxwell and Waterston. During the nineteenth and early twentieth centuries it reached a high degree of sophistication as a result of the contributions of a very large number of workers, and it can be said that the equilibrium and transport properties of a gas, at not too high densities, can be predicted quantitatively if the interaction between molecules is regarded as known. Alternatively, the variation of these quantities with density and temperature can give quantitative information about this interaction function.

The kinetic theory of solids was launched in the early twentieth century by various workers, including Einstein, Nernst, Lindemann, Debye and Born.

Simplifications are possible here for reasons just exactly opposite to those which permitted progress in the case of a gas. Whereas a gas is a completely disordered structure in which a typical molecule spends most of its time running free between collisions, a crystalline solid has an ordered repeating lattice structure with only occasional empty lattice sites, misplaced molecules and other breaks in the regularity. Furthermore, a typical molecule will spend a comparatively long time oscillating about the same equilibrium position in the lattice, migrating to another lattice site only occasionally. We can study an assembly of coupled oscillators by using the generalisation of mechanics due to Lagrange and others, while at low temperatures it becomes necessary to replace classical mechanics by quantum mechanics. It turns out that we can think of a crystalline solid in two apparently quite different ways, which turn out to be equivalent. We can think of it as an assembly of coupled oscillators, or we can describe the motion of the molecules as being equivalent to elastic waves of all possible frequencies traversing the solid in all directions. These elastic waves turn out to be subject to a law quite similar to Planck's law governing radiation in an enclosure, namely, that an oscillator of frequency v can only gain or lose energy in units of hv, where h is Planck's constant. This is sometimes expressed by saying that elastic or sound waves are quantised. By analogy with light quanta, sound quanta are often known as 'phonons'. This study of the mechanical and thermal properties of solids together with an attempt to relate them to the interactions of individual molecules is one branch of solid state physics. Other branches are concerned with the interpretation of the magnetic and electrical properties of solids in terms of those of the individual molecules and electrons.

The history of the kinetic theory of liquids is one of a number of fundamental advances, often separated by quite long intervals. It had long been realised that a liquid is unable to resist shearing stresses permanently because of its disordered structure. It would be expected that a solid, with nearly perfect lattice structure, would resist shear. This is because such a configuration is likely to be one in which the potential energy of a typical molecule is nearly a minimum value, in other words work is required to displace it from its lattice site. If we have a periodic lattice structure with only occasional imperfections, neighbouring rows of molecules would be changed in relative position by a shear, and this can be shown to require work if the interaction between molecules is like that shown in Fig. 1.1. If, as in a liquid, there is no lattice structure, then a great many molecular configurations with very nearly the same potential energy exist which can be occupied in turn, and a shear deformation (which does not alter the density or the mean distance between the molecules) can take place relatively easily.

In 1850 Berthelot produced definite experimental evidence that liquids could withstand tensions of at least tens of atmospheres. This may be taken as final confirmation of the fact that the forces between molecules are attractive at large distances, in other words that a liquid resists changes of density (that is of mean distance between the molecules) even though it cannot resist changes of shape. Berthelot also found that, as one might expect, the extensibility of a liquid was numerically about equal to its compressibility. In this respect a liquid resembles a solid and is in sharp contrast with a gas, which expands to fill any space, however large, to which it has access.

In 1873 van der Waals proposed his famous equation of state, which allows qualitatively both for the finite size of the molecules and for their mutual attraction. Since the attraction is expected to become more important as the density decreases, the prediction of theory is that at any temperature below a certain critical value T_c compression of an imperfect gas to a certain density is followed by a 'landslide'; the assembly collapses to a larger density at which the molecules are nearly in contact. We identify this with condensation to a liquid. These simple corrections to the 'perfect gas' laws are capable of accounting qualitatively not only for the observed departure of real gases from perfection, but for the observed phenomena of condensation, the critical temperature and even some of the properties of the liquid phase as well.

Little further progress was possible until something definite was known about the structure of a liquid. After von Laue and the Braggs discovered that crystalline structure could be determined by X-ray diffraction, it was natural to examine liquids by X-rays. In consequence the introduction of the concept of the **distribution function** of a liquid followed in the 1920's. This quanitity is the variation of local density that would be measured by an observer at the centre of a typical molecule. No other molecule can approach very much closer than the gas kinetic diameter but at this distance we can surround a rigid sphere by up to twelve others. We can conclude that a typical molecule is surrounded by 'Chinese boxes' formed by the other molecules in the liquid. This means that, as a function of distance from the central molecule, the local density oscillates in the manner shown in Fig. 1.2 approaching a constant value at large distances. The interest of this quantity, usually denoted by $g(r)$, is two-fold. In the first place, it can be measured experimentally, since it can be determined by X-ray or neutron scattering. Secondly, the value of $g(r)$ is proportional to the number of atom pairs separated by a distance r while the total potential

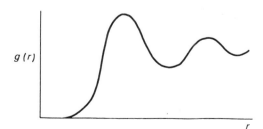

Fig. 1.2 The distribution function of a typical liquid.

energy of the liquid is proportional to the integral $\int r^2 \phi(r) g(r) dr$, where $\phi(r)$ is the interaction energy between a pair of molecules with centres at a distance r. Thus if $g(r)$ is known as a function of temperature and density (and if the interaction energy is entirely between pairs of atoms) the internal energy is also known as a function of temperature and density, and therefore all the equilibrium properties of the liquid can be deduced thermodynamically.

Thus it is of great interest to ask what is the relationship between $g(r)$ and the interaction function $\phi(r)$. For a gas of very low density, we should have simply that $g(r)$ was proportional to the Boltzmann factor $\exp(-\phi(r)/kT)$. For a liquid, we must allow for multiple encounters between molecules, and in 1935 Kirkwood showed that the proper mathematical tool for this was an **integral equation** relating $g(r)$ to $\phi(r)$. This equation has never been solved exactly, but Kirkwood and his students were able to solve various approximate versions of it. These amounted to the first *a priori* calculations of the equation of state of a liquid. These workers also showed that for the 'rigid sphere' type of interaction ($\phi(r)$ infinite for r less than the sphere diameter and zero otherwise), the mathematical form of $g(r)$ should change at a certain density. This, they suggested, was related to the solidification transition – a conclusion that ultimately turned out to be correct, though many people found it difficult to accept the idea that purely repulsive interactions could bring about the appearance of an ordered structure like a crystal lattice. This work was later repeated on slightly different lines by Born and Green, who reached very similar conclusions.

In 1946 Kirkwood showed that the transport coefficients of an imperfect gas or liquid could also, in principle, be related to $g(r)$ and to the modifications in this quantity caused by gradients of density, temperature or velocity. These ideas are still under development; unlike the statistical mechanics of equilibrium processes, the statistical mechanical theory of non-equilibrium processes is still incomplete, and some difficulties of principle remain.

No definite conclusions could be drawn merely by using the Kirkwood or Born-Green theories to calculate equilibrium properties of a liquid like argon and then 'comparing them with experiment'. In the first place the function describing the interaction of two atoms even as simple as argon was imperfectly known. In the second place the interaction between argon atoms is probably not strictly of two-body type. (The interaction energy of two atoms may easily be affected by the proximity of a third because of the resulting distortion of the outer electron shells). Indeed, recent theoretical and experimental work clearly shows that three-atom interaction effects *are* important. Thus the task of 'comparing theory with experiment' is rather difficult even if we discounted altogether the approximations that always have to be made when using the integral equation approach. The effect of some of these is extremely hard to assess numerically.

Thus no definite check on the integral equation approach was possible until

1957. In that year independent calculations were made on the hypothetical assembly, the rigid sphere gas, by the large computers at Livermore and Los Alamos. The Livermore computations were made by actually solving the equations of motion for small assemblies of tens or hundreds of particles, the Los Alamos computations by calculating the equation of state of such an assembly by a 'Monte Carlo' process, that is to say by a systematic sampling of all the permissible configurations of the assembly. The results of these calculations, using two quite independent methods, agreed well. They confirmed the predictions from the Kirkwood and Born–Green integral equations that a transition to an ordered, solid-like configuration should occur when the assembly has a volume of about 1.5 times the close-packed volume. However, the numerical calculations of the pressure of this hypothetical assembly of rigid spheres disagreed badly at liquid densities, the pressures calculated from the integral equations being only about a third of those obtained from the machine calculations. Consequently one had to conclude that the integral approach could only be relied upon for 'order of magnitude' results at liquid densities. Thus it did not seem to be much of an advance on the van der Waals approach, which also gives orders of magnitude.

In 1958 a new integral equation was introduced by Percus and Yevick. Although it is also based on a number of approximations they are quite different from those used by Kirkwood and Born and Green, and the pressures calcuated for the rigid sphere and other reasonable models agree with the machine calculations within a few per cent. The reason for this much better performance of the Percus-Yevick and various later integral equations is still not clear, as there seems at first sight little to choose *a priori* between the old and the new types of approximation. This matter is under investigation, for it is obviously of crucial importance.

Meanwhile there has been intensive study of the nature of the critical region. It is now known that the description of it given by the work of van der Waals and similar equations of state is very rough. For example, the van der Waals theory predicts that below the critical temperature the boundary of the two-phase region should be described by the relation $|T_c - T| \propto |V_c - V|^2$, whereas experimentally the relationship seems to be cubic rather than quadratic. The variations of a number of other quantities (e.g. specific heat, compressibility, coefficient of expansion) have also been studied. In all cases the variations differ significantly from what would be predicted from the van der Waals equation. (We are assuming that the objective of theory is to account for the various power laws representing the rapid variations of observed quantities in the immediate neighbourhood of the critical point.)

It can be said with confidence that the main cause of these discrepancies has been traced. From the work of Uhlenbeck and others it is known that an equation of van der Waals type would be the rigorous consequence of an attractive force of extremely long range. However the real interaction energy is inverse sixth power, and therefore becomes neligible within a distance of three or four gas-kinetic diameters.

The same type of problem arises with the magnetic properties of solids. Although these are 'lattice' rather than 'continuum' problems, they are believed to be mathematically similar. For such models the effect of a very long range interaction is precisely known; it amounts to the Weiss or 'molecular field' theory of ferromagnetism and antiferromagnetism. The opposite extreme case, in which interaction between the atomic magnets is supposed to be confined to nearest neighbours on the lattice, has been studied in great detail by Domb and his co-workers. The verdict is that the 'critical exponents' (describing the laws of variation of the various quantities near the critical point) are indeed expected to be quite different for long and short range interactions, and this conclusion is now supported by an impressive body of experimental evidence. It is possible that the critical exponents depend much more on the number of dimensions (they are always quite different for two-dimensional and three-dimensional models) than they do on the lattice structure. Thus we might expect them to be nearly the same for 'lattice' as for 'continuum' models, and there is also evidence for this. For a discussion of this evidence the reader is referrred to the Proceedings of the 1965 Conference on Critical Phenomena at Washington.

The quantitative theory of a liquid based on the two-molecule distribution function has almost entirely replaced earlier models. Examples of these are the 'hole' model, in which the liquid is thought of as a crystal lattice with an appreciable percentage of the lattice sites unoccupied, and the 'free-volume' or 'cell' model, in which each molecule is supposed to move in a smoothed potential field representing its average interaction with its neighbours. Various attempts were also made to compromise between a solid and a gas, such as the suggestion that a liquid consists of an equilibrium distribution of 'crystallites' of all sizes from a few molecules upwards. There was also the 'significant structure' postulate, according to which a typical molecule spends a certain proportion of its time in a 'gas-like environment'. All of these models contain adjustable constants that are difficult to calculate *a priori*. In practice they fared very similarly to the original van der Waals theory: suitable choices of the constants gave agreement with some of the data, but other experimental quantities were predicted in order of magnitude only.

All these tentative models can be regarded as first approximations to the true situation described by the distribution function. However, they retain their utility becasue they give physical insight using fairly simple mathematics. The real difficulty is that the distribution function $g(r)$ never differs very much from that appropriate to rigid spheres; in fact the changes in $g(r)$ produced by a change in the interaction function $\phi(r)$ are always rather small. However the calculated value of a quantity like pressure does turn out to be very sensitive to small changes in $g(r)$ and the calculated values of transport coefficients are more sensitive still. That is to say, the theory relating $g(r)$ to $\phi(r)$ must be considerably more accurate, and the approximations must be such as to produce errors in $g(r)$ itself considerably less than the 5 -10% that we would regard as acceptable agreement between 'calculated' and 'observed' equations of state. An appreciable contribution to the calculated pressure comes from the portion of the curve for $g(r)$ for which r is only slightly greater than the gas-kinetic diameter. Unfortunately this is the very portion of the curve that is most difficult to predict theoretically! Also, since it is related to scattering at high angles, this portion of the curve for $g(r)$ is very difficult to deduce from X-ray or neutron scattering!

Theories involving the distribution function are more fundamental than those involving 'liquid models', but for the above reasons they call for a great deal of computational effort.

References

Nat. Bureau of Standards Miscellaneous Publication 273, *Critical Phenomena, Proceedings of a Conference, Washington D.C.*, 1965.

Daniel Bernoulli
1700–1782

One of a distinguished family of talented mathematicians whose magnum opus, *Hydrodynamics,* which appeared in 1738, laid the foundations of hydrodynamics. (By courtesy of Dover Publications, Inc., New York)

**J. D. Van der Waals
1837–1923**

His famous equation was published in 1873 in his dissertation
for his doctor's degree.

Marcelin Berthelot

at work in his laboratory. This
famous French scientist's pioneer experiments in 1850
showed that it was possible for
liquids to withstand tensions of
at least tens of atmospheres.
(Photograph reproduced by
courtesy of the Editor of Nature)

Chapter 2
APPROACHES TO THE LIQUID STATE

2.1 THE THREE PHASES OF MATTER

We have described these from the molecular point of view in Chapter 1. The three types of molecular arrangement can be roughly illustrated by the following analogy. Let us imagine a number of people. They can be sitting in the regularly arranged seats of a modern lecture theatre (solid) or they can be talking in close groups and moving slowly around as in a crowded cocktail party (liquid) or they can resemble soccer players dashing around in a large field (gas). Now, consider the effect of temperature on matter. At very low temperatures all matter (except helium at low pressure) is solid, at very high temperatures (as in some stars) it is gaseous. However, at intermediate temperatures a large proportion of such matter exists as liquid. So, as in the molecular point of view, the intermediate position of the liquid is again exemplified. Such considerations have played an immense part in our understanding of liquids and have formed the basis of two of the main approaches to the liquid state. These assume a liquid either to be a highly compressed gas or to be a loosely-bound crystalline solid. These two approaches will be discussed in Sections 2.2 and 2.3.

Another familiar experience is that matter can exhibit changes of phase. A solid can melt and change into a liquid; a liquid can boil and change into a vapour, can become a gas. Thus we have two important temperatures in ordinary experience: the melting or freezing temperature and the boiling point, but we must be careful not to ascribe too much importance to these temperatures for the following reason. Taking water as our example, this liquid boils at $100°C$ because its vapour pressure at this temperature happens to be equal to the pressure of the earth's atmosphere. If we lived on another planet this boiling point would be different because of the different atmospheric pressure. Thus the boiling point is a property not only of the liquid but also of the ambient atmospheric pressure and its rate of change with such pressure is quite big. Similarly the melting point varies with the ambient pressure, though not at such a rapid rate. These considerations suggest that we should isolate a specimen of matter, be it solid, liquid or gas, from its surroundings and enquire whether there are other temperatures of a more fundamental nature than the boiling and melting points.

To do this we enclose our specimen of matter in a cylinder with a movable piston, and ensure that the air inside has all been previously removed. If the cylinder, whose walls are assumed to be transparent, contains a liquid and its saturated vapour as in Fig. 2.1 then the rate at which molecules leave the liquid to enter the space above (evaporation) is equal to the rate at which the vapour molecules return to the liquid (condensation). This dynamic two-way transfer is said to be a state of thermodynamic equilibrium.

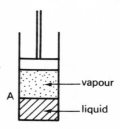

Fig. 2.1   Cylinder, with movable piston, containing a liquid and its vapour.

This simple apparatus can be used to investigate how the behaviour of the contents of the cylinder varies with pressure, volume and temperature. In our search to find some meaningful temperature to replace the ordinary boiling point, let us suppose that we have water and water vapour at 15°C in our cylinder as in Fig. 2.1. The vapour and liquid regions will be clearly separated by the boundary surface A. The liquid density ρ_l will be just less than 1000 kg m^{-3}, the vapour density ρ_v about 1.37×10^{-2} kg m^{-3} and the vapour pressure $P = 0.0168$ atm. Next suppose that the temperature is raised, the volume being kept constant by ensuring that the piston is maintained in the same position throughout. As the temperature rises the liquid will expand and therefore decrease in density while more molecules will leave the liquid and enter the vapour phase, thus causing the vapour density and also the vapour pressure P to increase. When the temperature reaches 100°C there will be no boiling observed because the contents of the cylinder are not exposed to the outside atmosphere, but P still attains the value of 1 atm at this temperature. As the temperature rises further ρ_l decreases and ρ_v increases, with the boundary surface still clearly visible. However, when the temperature 374.15°C is reached, ρ_l and ρ_v become equal. At the same time the dividing boundary A between liquid and vapour disappears and it is no longer possible to observe two different regions within the cylinder. As the temperature is raised still further this uniform nature of the contents of the cylinder persists and we say that the cylinder now contains a *fluid* which is neither liquid nor gas. The temperature T_c at which the liquid-vapour boundary disappears is known as the **critical temperature.** It is of fundamental importance because it is the upper temperature limit at which the liquid phase can exist. At temperatures above T_c all the properties which distinguish a liquid from a gas disappear and we simply have 'fluid-like' properties. Furthermore, changing the pressure by moving the piston does not change the situation and does not result in a separation into liquid and gaseous phases. The critical data for water are $T_c = 374.15°C, P_c = 218.3$ atm and $\rho_c = 320$ kg m^{-3}.

Next let us seek a temperature which is a natural lower limit to the liquid phase in equilibrium. If the contents of our cylinder are now cooled at constant volume to below T_c, the original liquid-vapour mixture reappears and eventually, at a certain temperature T_r, ice begins to form. If the temperature of the system is kept constant at this value T_r the solid, liquid and vapour can coexist in thermodynamic equilibrium, and for this reason the temperature T_r is called the **triple point.** As soon as the temperature falls below T_r the liquid disappears as it is converted into ice and a thermodynamic equilibrium exists between ice and vapour. Thus T_r is the temperature below which the liquid phase cannot exist in equilbrium with vapour and is therefore the lower limit which we anticipated. For water $T_r = 0.01°C$, a little above the 'normal' freezing point.

Experiments on the above lines, in which two or more phases of matter coexist in thermodynamic equilibrium in a closed container, were carried out by de la Tour (1822) and Andrews (1869). These will be discussed further in Section 2.2.

A word must now be said about matter *not* in thermodynamic equilibrium, that is, according to our present discussion, in an *open* container. We are all familiar with the strong odour of a moth-ball. The existence of this odour shows that a large number of its molecules are continually leaving it, and after a certain time the ever-dwindling moth-ball disappears completely. This disappearance of such a substance when left to itself occurs by the process known as **sublimation.** Again consider liquids. All liquids are in fact volatile to a greater or lesser degree, that is they evaporate completely after a time if in an open container. Some liquids, such as petrol, are highly volatile and evaporate very quickly.

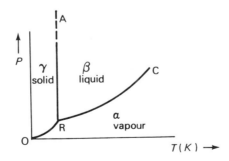

Fig. 2.2   Phase diagram for a simple substance.

Let us return to the cylinder with its piston and contents. A (P,T) or (P,V) diagram can be constructed to show the equilibrium between the three phases inside the cylinder. A (P,T) diagram (not necessarily shown to scale) for a simple substance, say a monatomic liquid like argon, is shown in Fig. 2.2 and from it we can make the following observations about the various curves shown.

The vapour pressure curve RC shows the (P,T) variation when the cylinder contains liquid and vapour only. R is the triple point, C the critical point, and this curve can have only the extent RC shown because for temperatures greater than the critical point and less than the triple point the liquid phase does not exist in equilibrium with vapour.

Curve OR shows the (P,T) variation when the cylinder contains only solid and vapour in equilibrium. Its lower limit is the point O where, at the absolute zero, the vapour has all condensed giving zero pressure and leaving only solid present. OR is called the **sublimation curve**.

Curve RA shows what happens when the cylinder contains solid and liquid only; in this case the vapour is all condensed by moving the lower end of the piston into contact with the liquid. RA is therefore the melting curve; it is steep, indeed almost vertical, and shows how the melting point varies with pressure. It must be emphasized that the point A is not the end of the curve for no such terminal point has been observed. In general the melting point rises with increasing pressure, but for water it is lowered and the portion RA slopes in a direction opposite to that shown in Fig. 2.2.

The point R is the one where the three regions (α), (β) and (γ) meet, that is, where the solid, liquid and vapour coexist in equilibrium.

2.2 THE 'GAS-LIKE' APPROACH TO LIQUIDS

To discuss the resemblance between a liquid and a gas it is best to start by reviewing our knowledge of the gaseous state.

The perfect or ideal gas is pictured as consisting of identical molecules each of negligibly small physical dimensions so that they are considered as points. These molecules move around in their enclosure and it is assumed that no intermolecular forces exist between them. The only interactions between molecules envisaged are those occurring when two molecules collide, but such collisions are comparatively rare. The total energy U of the gas is then equal to the sum of the kinetic energies of the individual molecules. There is no other contribution to U because, as there is no intermolecular force between any two molecules, there is no mutual potential energy between these two molecules either. For a perfect gas of N molecules the result turns out to be

$$U = \frac{3}{2}NkT \tag{2.1}$$

where k is known as Boltzmann's constant. For one mole of gas we can write

$$U = \frac{3}{2}RT. \tag{2.2}$$

In this case $R = Nk$ where N is equal to 6.02×10^{23} (Avogadro's number) and is the number of molecules in one mole of a gas. R is known as the universal gas constant and has a value of $8.314 \ \mathrm{JK^{-1} \ mol^{-1}}$. k is therefore the gas constant per molecule and has a value $(8.314/6.06) \times 10^{-23} = 1.38 \times 10^{-23} \ \mathrm{JK^{-1}}$.

On the basis of this model the kinetic theory of gases leads to the well-known equation of state for one mole of gas in the form

$$PV = RT. \tag{2.3}$$

If we consider the temperature of the gas to be kept constant then (2.3) reduces to Boyle's law, namely

$$PV = \text{constant} \tag{2.4}$$

and the corresponding variation of P with V at constant temperature is called the (P,V) isotherm.

If the density of the gas is low, that is, if the molecules are far apart for most of the time, we would expect the effect of intermolecular forces to be negligible. Thus for these low densities we would expect equations (2.3) and (2.4) to be satisfied by a real gas, and this is found to be so. If the density of the gas increases, the molecules will be closer together and we must enquire how this affects the intermolecular forces between the molecules. What can we say about such forces? The fact that there are cohesive forces present which hold the molecules together in liquids and solids suggests that forces between molecules are attractive. Yet if we try to compress an ordinary liquid or solid this is very difficult, suggesting that the force between two molecules becomes strongly repulsive below a certain molecular separation. To deal with this situation the molecules must be regarded not as mass points but rather as nearly 'rigid' spheres of diameter σ; the minimum possible distance between the centres of two molecules will then be σ (see Fig. 2.3). For example, in the case of liquid argon,

Fig. 2.3 Two molecules, regarded as 'rigid' spheres, at their minimum distance apart.

$\sigma = 3.4\text{Å}$. This idea of replacing 'point' molecules by rigid spheres of finite diameter σ is then a convenient way of saying that when the distance r between two molecules decreases to about σ, a very strong or infinite repulsion between them occurs. When r is greater than this, the force is an attraction which varies as r increases. If r becomes fairly large, say of the order of five molecular diameters as in a gas of low density, then this attraction is virtually zero.

The following simple calculations show that under 'standard' conditions (S.T.P.) 0°C and 1 atmosphere pressure, nearly all of a gas is empty space and the attractive forces between two typical molecules are negligible. A mole of gas at S.T.P. fills 22.4 litres and contains about 6×10^{23} molecules. Hence we allot a volume of 3.7×10^{-20} cm³ to each molecule, which corresponds to a cube of side about 3.3×10^{-7} cm. This would be the distance between two neighbouring molecules if they were arranged on a regular cubic lattice. A little reflection shows that, if the molecules were arranged at random, the average distance between nearest neighbours must be of this same order of magnitude. Compare this with the figure of about 5×10^{-8} cm for the molecular diameter, which is negligible from this point of view. Further (see Fig. 3.1), two molecules are on the average far enough apart for attractive forces to be negligible. Thus we can understand that, under standard conditions, the departure of most gases from 'perfection' is negligible. However, if we compress a gas to a density of the order of thousands of that corresponding to S.T.P., we should force the molecules practically into contact. At such densities either solidification or liquefaction usually occurs, which it is depending on whether we are outside the interval between the critical temperature and the triple point, or between the two.

The liquid phase can only exist in equilibrium with vapour if the pressure is between that corresponding to the triple point P_r, which is typically a fraction of an atmosphere, and P_c, the critical pressure, typically tens or hundreds of atmospheres. Therefore practically every substance can be brought into the liquid phase under conditions reasonably accessible in the laboratory. The only apparent exceptions are:

(a) Substances like iodine, which only have a very small liquid range of temperatures, and are practically always found as solid or vapour.

(b) Substances like carbon dioxide, which can exist as liquids only under pressure. Since P_r is greater than 1 atmosphere, carbon dioxide can only exist as a solid or vapour at atmospheric pressure (hence the production of 'dry ice' when the liquid is released from a cylinder).

(c) Substances like cellulose, plastics and some other organic compounds, which decompose if one tries to melt them.

(d) Substances like carbon and osmium, whose melting points are near the limit of what is readily attainable in the laboratory.

Triple-point and critical temperatures run from tens to thousands of degrees absolute for the vast majority of substances. It is relatively easy to take most substances above their critical temperatures in quite ordinary laboratory experiments, and far more extreme conditions exist in the universe, for example in the interiors of planets and stars. It is no accident that conditions on the earth correspond roughly with those for which the liquid phase is possible. All biological (and most meteorological) processes depend crucially on the existence of the liquid phase. Therefore we can be sure that any other planet supporting life in anything like the forms we know (even if it depended on some liquid other than water) would have to have conditions not too far removed from those of the surface of the earth. Consequently one of the key points in the study of liquids is the search for relationships between the intermolecular interactions and the pressures, densities and temperatures corresponding to the triple and critical points.

One of the outstanding discoveries in the field was due to van der Waals, and it is still of importance today. He showed that the critical point arises from an approximate balance between the effects of the short-range repulsive forces and the longer-range but weaker attractive forces. He represented the repulsion simply by replacing the 'point' molecules of a perfect gas by rigid spheres of finite diameter σ. If V is the volume of an enclosure occupied by a gas, then the total available volume at any instant is not V, as for point molecules, but something less because of the finite volume of the molecules. Thus V is replaced by $(V-b)$ where, as we shall show later, b turns out to be four times the total volume of the N molecules in the enclosure. Thus $b = \dfrac{2}{3} N\pi\sigma^3$. The attraction is dealt with by assuming that such attractions between the molecules are equivalent to some internal pressure P' causing the molecules to crowd closer together and that this pressure P' must be added to the applied pressure P. Van der Waals further suggested that P' is proportional to the square of the density of the gas, which means that $P' \propto 1/V^2$ for a constant mass of gas. Thus equation (2.3) is modified to

$$\left(P + \frac{a}{V^2} \right) (V-b) = RT \qquad (2.5)$$

which is the usual form of van der Waals's equation.

By giving T various fixed values and using equation (2.5) we can draw a series of (P,V) isotherms. These are shown in Fig. 2.4. Consider first the curve for $T = T_1$. It shows a minimum A_1 and a maximum B_1. The portion $A_1 B_1$ where the volume increases as the pressure is also increased (that is, dP/dV is positive) does not 'make sense' physically and we shall discuss this later. At a higher temperature $T = T_2$ the curve has the same general shape, except that the corresponding minimum and maximum, A_2 and B_2, have moved closer together.

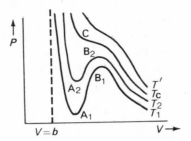

Fig. 2.4   Van der Waals isotherms

As T rises above T_2 the corresponding turning points move still closer together until at some temperature $T = T_c$ they merge into one point C, which is a point of inflexion. For temperatures $T' > T_c$, P decreases continuously as V increases. All the above is what a mere mathematical plotting of equation (2.5) tells us. How does this compare with what occurs in practice? To answer this we return to our cylinder and piston experiment, with water as our substance.

Let us consider a temperature $T_1 = 15°C$ and suppose that our piston is drawn well out so that the cylinder contains only water vapour. Then the point D_1 in Fig. 2.5 will represent this initial state. Next let us move the piston in,

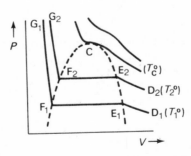

Fig. 2.5 (P, V) isotherms obtained by experiment

thereby decreasing the volume, but keeping the temperature fixed at T_1 throughout. It is then found that the isotherm follows the path $D_1 E_1$ where the pressure at E_1 is 0.017 atm. At E_1 however, drops of water will start to condense on the inside walls and piston base, and as the volume is further reduced more and more of the water condenses although the pressure inside remains constant. Thus the curve to the left from E_1 follows a horizontal path until we reach the point F_1. Here *all* the vapour will have condensed and the piston will now rest on the face of the liquid itself. Thus the region $E_1 F_1$ is a region of liquid-vapour equilibrium. If the piston is now pushed in further we shall be compressing the liquid, and this requires a very great change in pressure to cause even a small volume change; this is shown by the steep rise $F_1 G_1$.

Suppose now that the experiment is repeated at a higher temperature $T_2 = 100°C$. Starting at D_2 the curve follows the path $D_2 E_2$ to E_2 at which condensation starts; at E_2, $P = 1$ atm. By further reducing the volume the remaining portion $E_2 F_2 G_2$ is traced out. In this case the horizontal portion $E_2 F_2$ is less than $E_1 F_1$; furthermore the volume at E_2 is less than that at E_1 and that at F_2 is greater than that at F_1. As T is raised beyond T_2 the extremities of the horizontal portions move closer together until at a certain temperature T_c they coincide at the same point C. This is the critical point we discussed earlier.

How do we reconcile the experimental curves thus obtained with the van der Waals isotherms plotted in Fig. 2.4? To do this it is best to consider one typical curve only; say the one for $T = T_1$. We start with the van der Waals isotherm for this temperature T_1, see Fig. 2.6, and then draw the horizontal line $F_1 E_1$ so that the two shaded areas shown are equal. The reason for this will be given in Appendix 1.

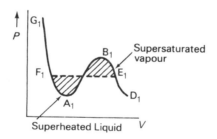

Fig. 2.6   Reconciling the curves of Figs. 2.4 and 2.5

We can then see the similarity between the above curve and the experimental isotherm corresponding to T_1 in Fig. 2.5. The portions $D_1 E_1$ in both cases represent the vapour region, the horizontal portions $E_1 F_1$ the liquid-vapour equilibrium, and the portions $F_1 G_1$ the liquid region. As we have already mentioned, the portion $A_1 B_1$ (where dP/dV is positive) is not physically realisable, but what of the regions $F_1 A_1$ and $E_1 B_1$ where dP/dV is still negative? Can these regions be realised in practice? The answer is 'yes' under certain conditions, and for this reason these regions are known as **metastable** regions.

Consider the region $F_1 A_1$ first. As we move down the region $G_1 F_1$ of the curve we would expect the liquid to start to evaporate at the point F_1 and follow the path $F_1 E_1$. However, it is possible for the van der Waals curve to be followed for some distance along the metastable region $F_1 A_1$, and whilst in this metastable state the liquid is said to be 'superheated'. This is the basis of the bubble chamber in which liquid hydrogen is subjected to an external applied pressure to prevent it boiling. If this pressure is suddenly released and if an ionising particle happens to pass through the chamber at the same instant, then the path of this particle will appear as a row of tiny vapour bubbles. The explanation is that the ionising particle acts as a nucleus for the initial growth of bubbles.

The possible existence of the region $F_1 A_1$ is also the basis of the famous experiments of Berthelot in 1850 on the stretching of liquids, as stated in Chapter 8.

Next consider the region $E_1 B_1$. As we move along the curve from D_1 to E_1 we expect condensation to set in at E_1. If, however, there are no suitable 'condensation' nuclei present the curve can extend for at least some way along the portion $E_1 B_1$ and the vapour in this metastable state is said to be 'supersaturated'. If nuclei (such as ionising particles) are present, condensation on them can take place, this fact is the basis of the Wilson cloud chamber.

As the temperature is raised above T_1 the points like F_1, A_1, B_1 and E_1 all move closer together until at $T = T_c$ these four points all merge into the single point C of Fig. 2.5 – the point of inflexion on the T_c isotherm. Thus the curve in Fig. 2.4 corresponding to the temperature T_c must be identified with the critical experimental isotherm of Fig. 2.5.

We shall return later to discuss the importance of van der Waals's equation, but enough has been said to show that there is a close relation between the liquid and gaseous phases. In particular we have seen that a liquid can be regarded as a highly compressed gas. This is based on the fact that, in the critical region, the density of a fluid can be increased continuously from a low value appropriate to a gas to a high value approaching that of a liquid. Our discussion has been based on the relatively simple interaction forces which are implicit in the van der Waals equation, but the conclusions will still hold for more realistic interaction forces.

2.3 THE 'SOLID-LIKE' APPROACH TO LIQUIDS

Whereas the similarity between a liquid and a gas was evident at temperatures near the critical, that between a liquid and a solid is more evident when we consider the transition from solid to liquid at the triple point (or, which is virtually the same thing, the melting point).

Consider the molecular arrangement in a crystalline solid. At the absolute zero the atoms or molecules in such a solid are nearly at rest at certain positions known as lattice points or sites. These sites form a definite regular geometrical pattern, known as a lattice, which extends throughout the whole interior of the solid right up to its surface. For example, in a simple cubic lattice the sites would be situated at the eight corners of identical cubes extending throughout the whole volume of the solid. Since this regular geometrical arrangement, or 'order', exists throughout the whole inside region the solid is said to possess long-range order. If we then choose any internal molecule as our reference the position of any other molecule, however far removed, will bear some definite correlation to our initial reference molecule. Any molecule will also have a certain number, z, of nearest neighbours on its adjacent sites; z is known as the coordination number of the lattice. These z nearest neighbours of a given molecule A will lie on a sphere of centre A; the next-nearest neighbours will lie on a second concentric sphere of larger radius, and so on. These concentric spherical surfaces are often referred to as 'shells'. When we heat up the solid above the absolute zero the molecules vibrate more and more as the temperature rises, but these vibrations still occur about a well-defined system of lattice sites so long-range order is still preserved. Some examples of lattice arrangements in various crystalline solids are the following:

(a) The alkali metals, in which the arrangement is a body-centred cubic lattice with one atom at a cube centre surrounded by eight others situated at the corners of the cube. Here $z = 8$.

(b) The inert gases, helium excepted, in which $z = 12$. Here we have atoms at the eight corners of a cube and six others each at the centre of the six cube faces; they form a face-centred cubic lattice.

Now consider what happens as the temperature of the solid is raised still further. With this increase of temperature a struggle occurs between the ordering influence of the intermolecular forces and the disordering influence of the increasing thermal motion of the molecules. When the solid melts into a liquid the original long-range order is destroyed, but a good deal of local or short-range order remains. By this we mean that the molecular arrangement near a chosen central molecule in the liquid will be a fairly regular one, not very unlike that in the corresponding solid. On the other hand when we move out to distances of 20 to 30 Å or so then the positions of the molecules there will bear no real spatial geometrical relations to the position of the chosen 'central' molecule. In other words there is local or short-range order just around any given molecule but no long-range order throughout the liquid. We shall see in a moment how to deal with this situation in a liquid by means of the radial distribution function.

At the melting point there is usually a volume increase of 5 to 15 per cent, indicating that the molecules in a liquid possess more 'elbow room' than those in the solid. Experiments based on X-ray and neutron diffraction show that there is a break-down of the long-range order of the solid at the melting point although local order remains. There is also evidence that the volume increase on melting is related not so much to an increase in distance between neighbouring molecules as to a general decrease in the coordination number z. For example, if z were twelve in the solid state then it might on average be reduced to nine in the liquid state; in other words each molecule in the liquid can be regarded as being surrounded by nine nearest neighbours and three empty sites or 'holes'. Indeed, such ideas form the basis of the 'hole' theory of liquids as suggested by Eyring and others (Section 4.4).

We now consider how the degree of local order in a liquid can be described quantitively. This is done by means of the pair distribution function or the radial distribution function. The former function is based on the probability of pairs of molecules occurring in the liquid and depends, in general, on the distance r_{ij} between two molecules i and j *and* on the orientation of this distance with respect to some specified direction. For spherically symmetrical molecules such as those of liquid argon, the pair distribution function depends only on r_{ij}, and for such a 'simple' monatomic liquid the pair distribution function is known as the radial distribution function. Confining our attention to this kind of liquid, let us take one molecule of centre O, and consider the surrounding environment as viewed by an observer at O. The number of molecules per unit volume at a distance r from O will be denoted by $\rho(r)$, the number density. As we move out along any straight line from O, $\rho(r)$ will vary with r; due to the nature of short-range order $\rho(r)$ will be sometimes more, sometimes less, than the mean number density ρ_0 throughout the liquid where .

$$\rho_0 = \frac{\text{Total number of molecules in the liquid}}{\text{Total volume of the liquid}} .$$

In other words the ratio $g(r) = \rho(r)/\rho_0$ can be greater than, less than, or equal to unity. The function $g(r)$ thus defined is the radial distribution function.

The graph of $g(r)$ against r is of the general shape shown in Fig. 2.7. The curve starts from nearly zero at the point A which corresponds to

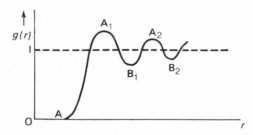

Fig. 2.7   The radial distribution function

the diameter of a molecule (a little thought will show that this must be so). For large r the value of $\rho(r)$ will tend to ρ_0, that is $g(r)$ tends to unity for large r. The departure of $g(r)$ from unity is then a measure of the short-range order in the liquid. The first maximum A_1 corresponds to the inner shell of nearest neighbours, A_2 to the second shell of neighbours, and so on; the minima correspond to distances intermediate to these shells. For example, for liquid argon at 84K, r is about 4.0Å at the first maximum and about 7.0Å at the second maximum.

It must be emphasized that both $g(r)$ and $\rho(r)$ for any value of r represent the mean values, averaged over time, at that distance; considerable statistical fluctuations from these values will occur due to the thermal movement of the molecules. The function $g(r)$ has been obtained from results involving the diffraction of X-rays and neutrons by liquids, and its importance in the theory of liquids will be discussed later.

In consequence we think of the properties of a liquid as being related to the relatively hard cores of real molecules, which are held together in a loose structure by attractive forces. It is therefore of great interest to study assemblies of macroscopic particles with analogous properties. Consider an ordinary powder of nearly uniform particle size. Under ordinary conditions it forms a heap needing a finite force to deform it. It can, however, be brought into a state very like that of a liquid. This is achieved by placing the powder in a vessel with a number of small holes in the bottom through which we can blow air or other gas. This process is known as **fluidisation** and since it is of considerable industrial importance it has been much studied recently. As we increase the rate of flow of the gas, the average spacing of the particles of powder increases, so that it expands almost uniformly. This is because part of the weight of each particle is supported by the upward stream of gas. If we increase the velocity further, each solid particle gains more 'elbow-room' and can then migrate through the bed for considerable distances. A bed of solid particles in this state is said to be fluidised.

A fluidised bed is observed to have many properties in common with those of a liquid. A typical solid particle can migrate slowly through the bed, but makes many collisions with its neighbours, exactly like a liquid molecule. If the stream of gas is reduced, the particles collapse into contact with their neighbours; if it is made too vigorous, the solid particles are all blown away. However, the intermediate range of velocities is relatively easy to find and maintain. Like a liquid, a fluidised bed has a well-marked 'free surface', and immediately above it there is a 'vapour phase' of much lower density consisting of the particles that have escaped through the free surface. It is possible to do experiments showing a fluidised material being 'poured' from one part of the vessel to another or to show it 'finding its own level' in two connected branches of the vessel. Under certain conditions, bubbles of gas often form in a fluidised material and rise through the bed in a way very similar to bubbles of gas in a liquid. We conclude from this that the characteristic property of a liquid, namely that its density remains practically constant but that it is readily deformed in shape, is a direct consequence of the form of the interaction between molecules. Molecules, as we have already said, have a hard core and an attraction of somewhat longer range. Thus we can expect a fluidised bed to act in some respects as a model of an actual liquid, and it does in fact reproduce quite a number of its properties.

However, like all models and analogies this one cannot be pressed too far.

The mechanism by which the particles of a fluidised bed are partially confined, a balance between gravity and the upward forces exerted by the gas stream, bears little resemblance to the mutual attraction between the molecules of a liquid which keeps them partially confined. On the other hand the comparison does show that some of the descriptions of a liquid, like the 'cell' model described later in Chapter 4.4, do have a direct relation with reality. Attempts at more detailed comparisons run into difficulties, because the pressure and temperature of a liquid do not seem to correspond with any well-defined properties of a fluidised bed.

2.4 MORE DIRECT APPROACHES TO THE LIQUID STATE

In view of the discussion in the two previous sections we would expect the imperfect-gas theory to be more applicable near the critical point and the disordered-crystal theory to be more valid near the melting (triple) point. We would not expect either approach to be particularly applicable at temperatures between these two points, and we must therefore consider approaches made to tackle the liquid state directly using the methods of statistical mechanics and with some knowledge of the intermolecular forces.

The greater difficulty involved when applying the methods of statistical mechanics to a liquid rather than to a gas or solid may be seen from the following considerations. One of the fundamental quantities in statistical mechanics is the partition function (see Chapter 3.1). For the present it suffices to say that once this function for the assembly has been determined, the other equilibrium thermodynamic properties can be readily obtained. In an ideal gas with no intermolecular interactions, the total partition function can be easily obtained because the only effective energies are the translational kinetic energies of the molecules. Again, in a highly ordered crystalline solid where the translational kinetic energies of the molecules are negligible, the molecules vibrate about equilibrium positions under the influence of strong intermolecular forces. Hence it is possible to derive a partition function for this case without great difficulty. However the intermediate liquid state is much harder to cope with in this way; there is a great deal of translational motion of the molecules in a liquid, but the cohesive intermolecular forces are strong enough to form a condensed state. In Chapter 3 we discuss the origin of intermolecular forces, and in Chapter 4 various models of the liquid state will be described.

REFERENCES

Books

Pryde, J. A., (1966) *The Liquid State,* (Hutchinson University Library, London).
Barker, J. A., (1963) *Lattice Theories of the Liquid State,* (Pergamon Press).

Tabor, D., (1969) *Gases, Liquids and Solids,* (Penguin Books).

Cole, G. H. A., (1967) *An Introduction to the Statistical Theory of Classical Simple Dense Fluids,* (Pergamon Press).

Temperley, H. N. V., Rowlinson, J. S. and Rushbrooke, G. S. (Editors), (1968) *Physics of Simple Liquids,* (North-Holland Publishing Company, Amsterdam).

Egelstaff, P. A., (1967) *An Introduction to the Liquid State,* (Academic Press, London and New York).

Papers and Articles

Alder, B. J. and Wainwright, T. W., (1958), *Nuovo Cim.,* **9**, Suppl. 1, 166.

Tabor, D., (1964) *Contemp. Phys.* **6**, 112.

Rushbrooke, G. S., (1968) Chapter 2 in *Physics of Simple Liquids*, referred to above.

Wood, W. W., (1968) Chapter 5 in *Physics of Simple Liquids*, referred to above.

Alder, B. J. and Hoover, W. G., (1968) Chapter 4 in *Physics of Simple Liquids*, referred to above.

Chapter 3
INTERMOLECULAR FORCES IN LIQUIDS

3.1 THE PARTITION FUNCTION AND
THE USE OF STATISTICAL MECHANICS

Consider an assembly of N particles, each of mass m, in an enclosure of volume V. If we assume that the temperature T is sufficiently high (for example, ordinary room temperature) the methods of *classical* statistical mechanics can be applied. We begin by assuming that our assembly is a perfect gas consisting of N essentially independent particles. The partition function f for one particle of the assembly is then given by

$$f = \sum_i e^{-\epsilon_i/kT} \qquad (3.1)$$

where ϵ_1, ϵ_2, ... ϵ_i ... are the permissible energies for any particle of the assembly. To evaluate f we can use the form $\epsilon = \tfrac{1}{2}m(v_x^2 + v_y^2 + v_z^2)$ for the energy of a particle and use a 6-fold integration over phase space. Alternatively we can use the quantum-mechanical form

$$\epsilon = \frac{h^2}{8m}\left[\frac{p^2}{a^2} + \frac{q^2}{b^2} + \frac{r^2}{c^2}\right] \qquad (3.2)$$

where p, q and r are integers and a, b, and c are the sides of a rectangular box of volume $V = a\,b\,c$. Using either method, it is shown in books on statistical mechanics that

$$f = V\left[\frac{2\pi mkT}{h^2}\right]^{3/2} \qquad (3.3)$$

where k is Boltzmann's constant and h is Planck's constant. It must be emphasized that f is the partition function corresponding to the translational kinetic energy of the particle only, and the result (3.3) is valid whatever the shape of the enclosure V.

Next we suppose that our assembly is a liquid consisting of N identical molecules. In a liquid the interactions between the molecules must be taken into account in addition to the translational kinetic energies, and we now do this. Let the molecules be labelled 1 to N. The number of distinct pairs of molecules ij in this assembly of N molecules is $^{N}C_{2} = N(N-1)/2$. If the mutual potential energy of molecules i and j at distance r_{ij} apart is $\phi(r_{ij})$ and if we assume that the molecules interact in pairs only then the total potential energy Φ of the assembly will be

$$\Phi = \sum_{i<j} \phi(r_{ij}) \tag{3.4}$$

where $\sum_{i<j}$ implies that each $\phi(r_{ij})$ is not counted twice and the summation in (3.4) contains a total of $N(N-1)/2$ terms.

It can then be shown that the partition function Z for *the whole assembly* of N molecules is given by

$$Z = \left[\frac{2\pi mkT}{h^2} \right]^{3N/2} Q \tag{3.5}$$

where the factor $(2\pi mkT/h^2)^{3N/2}$ takes account of the kinetic energies of the molecules and Q is the **configurational partition function** given by

$$Q(N,V,T) = \frac{1}{N!} \int \dots \int \exp(-\Phi/kT) \, dx_1 \dots dz_N \tag{3.6}$$

where the integral is taken over $dx_i \, dy_i \, dz_i$ for all values of i from 1 to N. Q depends on the various values of the separations r_{ij} for all possible configurations of the molecules and describes the effect of their interactions on the equation of state and equilibrium thermal properties of *any* assembly.

Once the partition function Z in (3.5) is obtained the various thermo-dynamic properties can be obtained also. The most important of these quantities are given in equations (3.7) to (3.10) below. The Helmholtz free energy A is given by

$$A = -kT \ln Z \tag{3.7}$$

and the internal energy U of the assembly by

$$U = \frac{kT^2}{Z} \left(\frac{\partial Z}{\partial T} \right)_V. \tag{3.8}$$

The equation of state is

$$P = -\left(\frac{\partial A}{\partial V}\right)_T \tag{3.9}$$

and the entropy S is given by

$$S = -\left(\frac{\partial A}{\partial T}\right)_V. \tag{3.10}$$

The results (3.7) to (3.10) are well-known in statistical mechanics and for further study the reader is referred to Appendix 2. In order to apply these results it is necessary to use the value of Z given in (3.5) and this involves a knowledge of the configurational partition function Q. Indeed the whole crux of the matter is to find Q and to do this the function $\phi(r_{ij})$ for the liquid must be known. This function $\phi(r_{ij})$ is now considered.

3.2 THE GENERAL NATURE OF INTERMOLECULAR FORCES IN LIQUIDS

Let us assume that we have a liquid consisting of a number of identical symmetrical molecules. First let us consider two of these molecules, labelled i and j, and ignore the presence of the others. A force will be exerted on each of these two molecules by the other and we shall assume that this force $F(r)$ depends only on the distance r between their centres. When this force is attractive it is conventional to give it a negative sign and when repulsive a positive sign. The general shape of the $(F(r), r)$ curve is shown in Fig. 3.1a.

For small values of r, less than a certain value r_0, $F(r)$ is positive and represents a strong repulsion between the molecules. For $r > r_0$, $F(r)$ is negative and therefore represents an attraction; as r increases beyond r_0 this attraction increases to a maximum (as represented by the point A) and then decreases to zero as the two molecules move infinitely far apart.

We next consider the mutual potential energy $\phi(r)$ and its relation to $F(r)$. $\phi(r)$ is the potential energy of the two molecules due to the intermolecular force and is often called the pair potential function. To increase the distance r between the molecules by a small amount dr the work that would have to be done is $F(r)dr$, which can be equated to the loss in the potential energy, that is $F(r)dr = -d\phi(r)$ or

$$F(r) = -\frac{d\phi(r)}{dr}. \tag{3.11}$$

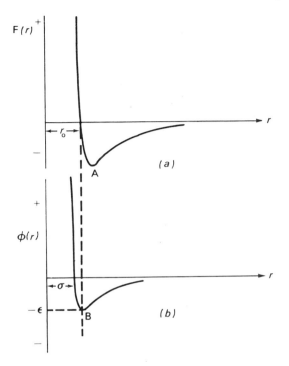

Fig. 3.1 Variation of (a) $F(r)$ and (b) $\phi(r)$ with separation between two molecules.

This relation enables us to obtain the $(\phi(r), r)$ curve shown in Fig. 3.1b and by convention we assume that $\phi(r) = 0$ when the molecules are infinitely far apart. This curve has a minimum at B when $r = r_0$.

Referring to Fig. 3.1a we see that at $r = r_0$ the two molecules are neither attracting nor repelling each other; a slight increase or decrease in r will cause an attraction or repulsion respectively. Thus $r = r_0$ represents the normal 'equilibrium' separation of the two molecules and r_0 is usually of the order of a few Ångstrom units. When the two molecules are at this equilibrium separation r_0 the potential energy $\phi(r)$ will be at its minimum value $-\epsilon$, the point B of Fig. 3.1b. If r is increased from r_0 to infinity the work that would have to be performed is numerically equal to ϵ, so that ϵ is often called the 'dissociation' energy. Functions $F(r)$ and $\phi(r)$ behave similarly as functions of r and we must always be quite clear which is referred to. In most of the literature $\phi(r)$ is used, because it appears explicitly in equation (3.6) while $F(r)$ does not.

For a physical interpretation to the value σ of r at which $\phi(r) = 0$ we consider two molecules to be initially at an infinite distance apart, in which case $\phi(r) = \phi(\infty) = 0$. Suppose that they now move towards each other along their line of separation with negligibly small velocities. Their initial total energy W_i, which is the sum of their potential and kinetic energies, is thus virtually zero. As the molecules approach each other changes occur in both the potential and kinetic energies, their sum remaining negligible. After being accelerated at first they eventually collide head-on. At the moment of collision their total kinetic energy is zero, so, if their total energy is still to be W_i, the value of $\phi(r)$ must also be zero. This occurs at $r = \sigma$ and so σ represents the effective diameter of each molecule because it is the closest distance which their centres attain in a gentle head-on collision (see Fig. 2.3).

The value of $\phi(r)$ at any value of r is really the algebraic sum of two energies, an attractive energy ϕ_A and a repulsive energy ϕ_R. The attractive energy between two molecules can be of various types. In a liquid it is usually of the type known as van der Waals binding, which is considered in Section 3.3 of this Chapter. The repulsive energy is considered in Section 3.4.

3.3 THE ATTRACTIVE ENERGY ϕ_A

3.3.1 Polar and non-polar molecules

Consider what happens when two atoms are bound together chemically to form a diatomic molecule. Some distortion of the original electron configurations of the atoms occur as a result and this may result in the molecule forming what is effectively an electric dipole. This may be explained as follows. The molecule as a whole is electrically neutral and may be considered to consist of n positive charges each of $+e$ (the protons) and n negative charges of $-e$ (the electrons). As far as electric effects *outside the molecule* are concerned, the n positive charges may be considered as being replaced by a single positive charge of $q = ne$ at some point A and the n negative charges by a single negative charge $-q$ at a point B. Thus the two charges at A and B may be considered to form an electric dipole of moment $\mu = qx$ where AB $= x$. Such a molecule, in which distance x is about 10^{-9} to 10^{-8} cm, is known as a **polar** molecule and possesses a **permanent electric dipole**. The unit in which μ is measured is known as the Debye unit. This is defined by putting $q = 10^{-10}$ e.s.u. (which is of the same order as the electronic charge 4.8×10^{-10} e.s.u.) and $x = 1\text{Å} = 10^{-8}$ cm. Thus one Debye unit is equivalent to a moment of 10^{-18} e.s.u. $\times$ cm. An example of a polar molecule for which $\mu = 1.08$ Debye units is the hydrogen chloride molecule.

If the two points A and B in a molecule coincide, as often happens, we have a **non-polar** molecule which does not possess a permanent dipole moment. Examples are O_2 and N_2 in which two *identical* atoms have been brought together and there is therefore no reason why the electronic configuration should be displaced towards one atom more than the other. Such symmetrical diatomic molecules consisting of two identical atoms are called homonuclear molecules and are non-polar. Conversely, diatomic molecules consisting of two different atoms such as HCℓ are heteronuclear and are very often polar.

Now let us consider more complicated molecules. If we have a symmetrical molecule, such as $C_,Cℓ_4$ in which we have four chlorine atoms symmetrically placed with respect to a central carbon atom, the molecule is non-polar. If one, two or three of these chlorine atoms are replaced by a hydrogen atom then we no longer have a symmetrical molecule and the molecule becomes a polar one. The water molecule is also weakly polar, because the centres of the two hydrogen atoms and the one oxygen atom are not in a straight line. The inert gases are monatomic elements and so they are obviously non-polar.

To sum up: all monatomic elements and all symmetrical molecules (whether homonuclear diatomic molecules or more complicated symmetrical ones) are non-polar. Nearly all remaining types are non-symmetrical and polar with a permanent dipole moment of μ_p, say.

Consider now what happens when some external electric field E is applied and suppose, for simplicity, that it acts along the axis of the dipole in the direction BA. In the case of a polar molecule the two charges will be subjected to forces in opposite directions along the axis and the separation between them will be increased by, say, an amount x'; this will cause an increase $\mu_i = qx'$ in the dipole moment, and μ_i is known as an induced dipole. In a non-polar molecule, A and B will be displaced by x' in opposite directions so that a dipole whose moment is $\mu_i = qx'$ is actually formed; in this case also μ_i is called an induced dipole. If the field is removed μ_i becomes zero: a polar molecule reverts to its original state of moment μ_p and a non-polar one to its state of zero moment.

The value of μ_i increases with the field and to a first approximation can be considered to be proportional to it, so that we can write

$$\mu_i = \alpha E \qquad (3.12)$$

where α is the polarisability of the molecule. In more complicated cases where the dipole axis makes an angle θ with the direction of field, a relation such as (3.12) still holds provided α is taken as being an average over all possible values of θ.

Let us consider the field produced by a polar molecule C of permanent moment μ_p at a point P at distance r from its centre, and let us suppose that a non-polar molecule D is situated at P (see Fig. 3.2). From electrostatics the

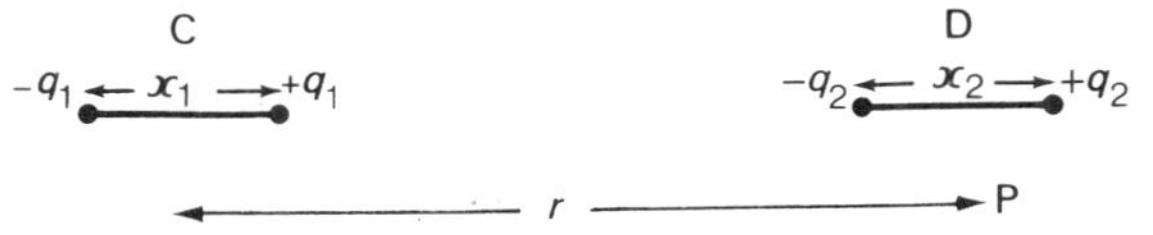

Fig. 3.2 A non-polar molecule, D, in the field of a polar molecule C.

field E at P due to the left-hand dipole C is $2\mu_p/r^3$ and this field* is directed from left to right; hence the charges $\pm q_2$ in the induced dipole formed in the second molecule D will be displaced as shown. The *attractive* force between the charges making up the dipoles has a magnitude

$$F_a = \frac{q_1 q_2}{[r - \frac{1}{2}(x_1 + x_2)]^2} + \frac{q_1 q_2}{[r + \frac{1}{2}(x_1 + x_2)]^2}$$

and the *repulsive* force a magnitude

$$F_r = \frac{q_1 q_2}{\left[r - \dfrac{x_1}{2} + \dfrac{x_2}{2}\right]^2} + \frac{q_1 q_2}{\left[r + \dfrac{x_1}{2} - \dfrac{x_2}{2}\right]^2}$$

and since $F_a > F_r$ the resultant force between the two molecules is an attraction.

If the dipole C in Fig. 3.2 is reversed end-to-end, the induced dipole in D is also reversed because the charge $+q_2$ is now displaced to the left and the charge $-q_2$ to the right; thus the resultant force between the two molecules is again an attraction. In fact, by an extension of this simple argument, it is not difficult to see that, when we have an electric field E produced by a polar molecule C, a second molecule D near to C at any particular orientation to it will always be *attracted* to this first molecule C. A similar example in magnetism is that of the attraction between a permanent magnet and a cylinder of soft iron; the induced moment in the soft iron is such that the iron cylinder is always attracted to the magnet whatever their relative orientations.

We are now in a position to discuss the mutual potential energy of attraction ϕ_A between two molecules. If we consider the most difficult case in which both molecules are polar we find that ϕ_A is made of three contributions as follows:

(1) The energy ϕ_s due to the mutual interaction between their two permanent dipoles.

(2) The energy ϕ_i due to the interaction of the dipole induced in each molecule with the permanent dipole of the other.

(3) A third contribution ϕ_L, known as the London or dispersion energy, whose origin will be discussed below.

We can then assume, without much error, that ϕs, ϕ_i and ϕ_L can be considered separately and added up to give ϕ_A

$$\text{i.e.} \quad \phi_A = \phi_s + \phi_i + \phi_L$$

---

*For simplicity, we shall not include the factor $4\pi\epsilon_0$ in the various electrostatic equations in this section.

3.3.2 The electrostatic attraction energy, ϕ_s

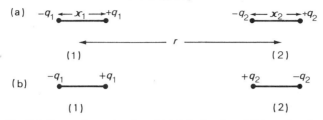

Fig. 3.3 Two polar molecules which (a) attract, and (b) repel each other.

Referring to Fig. 3.3a, let (1) and (2) be two **permanent** dipoles whose axes are in the same straight line. The distance between the dipole centres is r and the two moments are $\mu_1 = q_1 x_1$ and $\mu_2 = q_2 x_2$. In the arrangement of Fig. 3.3a the two dipoles attract each other and it is shown in textbooks on electricity that their mutual potential energy in this case is given by

$$\phi = -2\mu_1\mu_2/r^3 . \tag{3.13}$$

If dipole (2) is reversed as in Fig. 3.3b then the two dipoles repel each other and their mutual potential energy is $+2\mu_1\mu_2/r^3$, that is, of opposite sign.

Now consider all the other cases in which the axis of dipole (2) makes all possible angles with that of dipole (1). Suppose, in one such case, that θ is the angle between the two dipole axes; in this case let ϕ_θ be the mutual potential energy of the dipoles. Then if dipole 2 is reversed end-to-end the mutual potential energy will be $-\phi_\theta$. By this argument it might at first be thought that the average mutual potential energy between the two dipoles, when all possible orientations of the corresponding molecules are considered, turns out to be zero. However this is not so, because all the orientations do not have the same probability; this is taken into account by introducing a weighting factor $e^{-\phi/kT}$. In other words orientations of low total energy are more favoured than those of high energy. When such considerations are applied the average mutual potential energy ϕ_s taken over all possible orientations turns out to be given by

$$\phi_s = -\frac{2\mu_1^2\mu_2^2}{3kTr^6} .$$

Note that ϕ_s is *always* negative so that the force between the dipoles is an attraction. When the two molecules are identical then we have

$$\phi_s = -\frac{2\mu_p^4}{3kTr^6} . \tag{3.14}$$

In all this treatment terms higher than $1/r^6$ have been ignored but, to this order, ϕ_s is proportional to $1/r^6$. ϕ_s decreases as T increases, which is what we would expect intuitively. Finally it should be emphasized that ϕ_s arises when the two molecules considered are polar, that is, possess a permanent dipole.

3.3.3 The induction energy, ϕ_i

Fig. 3.4 Permanent dipole producing a field E at P.

Consider a permanent dipole μ_p and a point P on its axis as in Fig. 3.4. The field E at P is given by $E = 2\mu_p/r^3$ and if a second molecule is placed at P an induced dipole $\mu_i = \alpha E = 2\alpha\mu_p//r^3$ will be formed in it. As we saw in Section 3.3.2 above these permanent and induced dipoles will attract each other and their mutual potential energy of interaction will be given by

$$\phi = -2\mu_p\mu_i/r^3 \qquad (3.15)$$

(cf. Equation 3.13) which on substituting for μ_i gives us

$$\phi = -4\alpha\mu_p^2/r^6 \ .$$

This is an attractive energy.

If the dipole μ_p in Fig. 3.4 is reversed end-to-end, so also will the induced dipole μ_i at P be reversed, and their mutual potential energy will still be attractive. All other possible mutual orientations of the two dipoles can be considered by introducing a suitable weighting factor. In this way the final induction energy ϕ_i between two identical polar molecules turns out to be, to a good approximation,

$$\phi_i = -2\alpha\mu_p^2/r^6 \ . \qquad (3.16)$$

Note that ϕ_i is proportional to $1/r^6$ but is independent of the temperature. ϕ_i exists only if at least one of the molecules is polar.

3.3.4 The London or dispersion energy, ϕ_L

We have seen that the electrostatic energy ϕ_s exists between two molecules, both of which are polar, and that the induction energy ϕ_i exists between two molecules of which one, at least, must be polar.

We now consider a liquid consisting of non-polar molecules. At first sight one would not expect any attractive force to exist between two such molecules if arguments such as those in Sections 3.3.2 and 3.3.3 are used. Yet we know that such forces exist because of the cohesion of this type of liquid. They are called dispersion or London forces.

The discussion concerning ϕ_s and ϕ_i in the two previous Sections was based on the classical theory of electrostatics. We can explain the dispersion forces between two non-polar molecules (1) and (2) on classical grounds as follows. The electrons in these molecules are moving rapidly around the corresponding nuclei. At any instant the electrons in molecule (1) have some definite configuration so that molecule (1) has an instantaneous dipole moment even though, on average over a period of time, the electrons are distributed in a spherically symmetrical manner. So at this particular instant the instantaneous dipole in (1) induces a dipole in (2). This gives rise to an instantaneous force of attraction between them. As the instantaneous electron configurations in (1) vary so will this instantaneous attractive force. Hence the dispersion force is given by averaging this instantaneous attractive force over all the corresponding electron configurations in the molecule. On classical grounds such an explanation seems adequate, but quantum mechanical considerations are needed for a quantitative calculation.

The explanation of these dispersion forces was given by F. London (1930) and is based on the quantum mechanical treatment of a linear harmonic oscillator. We first concentrate on one electron moving around the nucleus of one atom and assume that its motion can be resolved into that of three linear simple harmonic oscillators along the x, y and z directions.

Let us consider, initially, the x-direction only. Hence our linear harmonic oscillator ˙consists of an electron, of charge $-e$ and mass m, vibrating about a point with a restoring force acting on it.

Consider first the classical treatment. The equation of motion of the oscillator is

$$m\ddot{x} + kx = 0.$$

At any instant the total mechanical energy W of the oscillator is

$$W = \tfrac{1}{2}kx^2 + \tfrac{1}{2}m\dot{x}^2,$$

where the first term on the right-hand side represents its potential energy and the second its kinetic energy. The frequency of the oscillator is given by

$$\nu_0 = \frac{1}{2\pi}\sqrt{\frac{k}{m}}. \tag{3.17}$$

When the methods of quantum mechanics are used the important difference is that the energy turns out to be quantized and to have possible values given by

$$E_n = (n + \tfrac{1}{2})h\nu_0$$

where ν_0 is given by (3.17) and $n = 0, 1, 2, 3, \ldots$ etc. The lowest energy value, corresponding to $n = 0$, is $\tfrac{1}{2}h\nu_0$, the 'zero-point' energy. This lowest energy state is the one in which the atom normally exists.

Now consider two such identical oscillators with their centres of oscillation at A_1 and A_2 on the x-axis as in Fig. 3.5. Let $A_1 A_2 = r$. If we now place a positive charge of $+e$ at A_1 and a similar charge at A_2, the two oscillators are

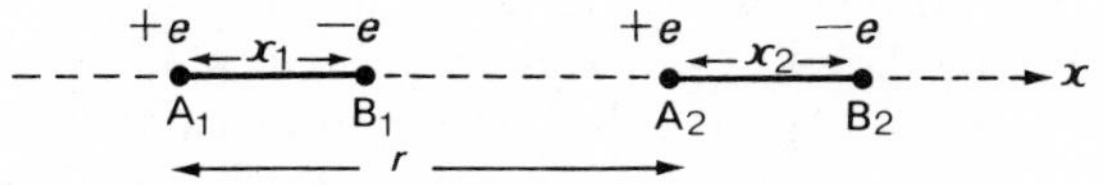

Fig. 3.5 Two fluctuating dipoles.

transformed into two dipoles whose moments fluctuate in value as the two electrons move back and forth about the points A_1 and A_2. The electrostatic interaction between these fluctuating dipoles is then essentially the same as the mechanism giving rise to the dispersion forces between two molecules.

The total energy W for the instantaneous state of affairs shown in Fig. 3.5 is made up of two contributions. First there is the total mechanical energy $\tfrac{1}{2}kx_1{}^2 + \tfrac{1}{2}m\dot{x}_1{}^2 + \tfrac{1}{2}kx_2{}^2 + \tfrac{1}{2}m\dot{x}_2{}^2$ and secondly the mutual potential energy $-2\mu_1\mu_2/r^3$ between the two dipoles, where $\mu_1 = ex_1$, $\mu_2 = ex_2$. This energy represents the instantaneous electrostatic interaction between the dipoles. Thus

$$W = \tfrac{1}{2}kx_1{}^2 + \tfrac{1}{2}m\dot{x}_1{}^2 + \tfrac{1}{2}kx_2{}^2 + \tfrac{1}{2}m\dot{x}_2{}^2 - \frac{2e^2 x_1 x_2}{r^3}. \tag{3.18}$$

For non-interacting dipoles the last term on the right-hand side would be absent. Its presence changes the natural frequencies of the oscillators and to deal with this situation we introduce two coordinates x' and x'' defined by

$$x' = \frac{x_1 + x_2}{\sqrt 2} \quad , \quad x'' = \frac{x_1 - x_2}{\sqrt 2}.$$

On substituting for x_1, x_2, $\dot{x}_1$ and $\dot{x}_2$ in (3.18) in terms of these new variables we readily obtain

$$W = \tfrac{1}{2}k_1 x'^2 + \tfrac{1}{2}m\dot{x}'^2 + \tfrac{1}{2}k_2 x''^2 + \tfrac{1}{2}m\dot{x}''^2 \tag{3.19}$$

where $k_1 = k - \dfrac{2e^2}{r^3}$

and $k_2 = k + \dfrac{2e^2}{r^3}$.

$$(3.20)$$

Equation (3.19) represents the energy of two non-interacting oscillators whose force constants k_1, k_2 are different and whose frequencies are

$$\nu_0' = \frac{1}{2\pi} \sqrt{\frac{k_1}{m}}$$

and

$$\nu_0'' = \frac{1}{2\pi} \sqrt{\frac{k_2}{m}} .$$

$$(3.21)$$

The total zero-point energy is now

$$E_0 = \tfrac{1}{2}h \left[\nu_0' + \nu_0''\right] .$$

$$(3.22)$$

If we use (3.20) and (3.21) to substitute for ν_0' and ν_0'' in (3.22) and expand as far as terms in $1/r^6$ we readily obtain

$$E_0 = h\nu_0 - \frac{h\nu_0 e^4}{2k^2 r^6}$$

which we will write as

$$E_0 = h\nu_0 - \frac{\beta}{r^6} .$$

Thus the total zero-point energy now is made up of two terms:
(a) The term $h\nu_0$ which is the total zero-point energy for two uncoupled oscillators, that is, oscillators without any interaction energy between them.
(b) The term $-\beta/r^6$, which represents the dispersive energy; this term is negative and therefore represents an energy of attraction proportional to $1/r^6$.

The method of finding the value of the dispersion energy between two actual molecules is then extended from the above one-dimensional treatment for one electron so as to include motion in three dimensions and also contributions from all the electrons in the molecule. The problem is a complicated one and the final value for the dispersion energy is found to be given by

$$\phi_L = -3\alpha^2 E_i/4r^6 .$$ (3.23)

In this equation α is the polarisability, which can be related to the force constant k and to the ionization energy E_i of the molecule. E_i can be put approximately equal to $h\nu_0$. The result (3.23) must be regarded as being a fairly good estimate for ϕ_L rather than an exact value. For fuller details the reader is referred to two papers by London (1930, 1937) and also to *Atomic Physics* by Max Born, Appendix (XL), (1969).

3.3.5 The resultant attractive energy, ϕ_A

We can now, to a good approximation, write the resultant attractive energy ϕ_A between two molecules as

$$\phi_A = \phi_s + \phi_i + \phi_L$$ (3.24)

since we assume that the three contributions on the right-hand side of this equation are regarded as mutually separate and independent. Strictly speaking both ϕ_s and ϕ_i should have been treated quantum mechanically, but the classical approach which we used in Sections 3.3.2 and 3.3.3 is adequate.

We note next that each of the quantities ϕ_s, ϕ_i and ϕ_L in (3.24) is proportional to $1/r^6$. Hence we may write

$$\phi_A = -A/r^6$$ (3.25)

where A is positive.

Table 3.1 Van der Waals interaction energies and other data for simple polar molecules at 20°C (values given by F. London).

Substance	$\mu_p \times 10^{18}$ (e.s.u.)	$\alpha \times 10^{24}$ (c.g.s. units)	$h\nu_0$ (eV)	$r^6 \phi_s \times 10^{60}$ (erg cm^6)	$r^6 \phi_i \times 10^{60}$ (erg cm^6)	$r^6 \phi_L \times 10^{60}$ (erg cm^6)
CO	0.12	1.99	14.3	0.0034	0.057	67.5
HI	0.38	5.40	12.0	0.3500	1.68	382.0
HBr	0.78	3.58	13.3	6.2	4.05	176.0
HCl	1.03	2.63	13.7	18.6	5.40	105.0
NH$_3$	1.50	2.21	16.0	84.0	10.0	93.0
H$_2$O	1.84	1.48	18.0	190.0	10.0	47.0

As far as the relative magnitudes of ϕ_s, ϕ_i and ϕ_L are concerned, it is true to say that ϕ_i is always much less than the others. ϕ_L nearly always makes the largest contribution to ϕ_A, and for non-polar molecules it is the *only* contribution to ϕ_A. Table 3.1 gives some examples of these various energies for some polar molecules.

ϕ_A is often referred to as the van der Waals interaction energy and $F_A = -\mathrm{d}\phi_A/\mathrm{d}r$ as the van der Waals force of attraction.

3.4   THE REPULSIVE ENERGY, ϕ_R

We have seen in Section 3.2 that when two molecules approach each other very closely a strong repulsion is set up between them. To obtain a value for this repulsive energy, even for the simplest molecules, is a very difficult problem. All we can do here is to give a brief qualitative discussion. As the two molecules get closer and closer together their electron clouds eventually overlap resulting in a strong repulsion between the molecules; this is because the clouds no longer completely shield electrostatically the two nuclei from each other. Various considerations lead to the conclusion that a repulsive energy of the form

$$\phi_R = \frac{B}{r^n},$$ (3.26)

where B and n are constants, is a reasonably good approximation.

An exponential form for ϕ_R is sometimes used instead of (3.26). This is more in accordance with quantum mechanics, according to which the charge density associated with the electrons falls off as $e^{-\lambda r}$ at distances comparable with the molecular radius. Therefore the overlap energy may be expected to behave in a similar way since it is an average of the electrostatic forces.

3.5   THE 12–6 ENERGY FUNCTION

The sum of ϕ_A and ϕ_R gives the *resultant* energy between two molecules. Thus, from (3.25) and (3.26)

$$\phi(r) = \frac{B}{r^n} - \frac{A}{r^6}.$$ (3.27)

In 1937, Lennard-Jones and Devonshire suggested an empirical form for the above energy with $n = 12$. This form is suitable for the energy between two non-polar molecules and for some weakly polar molecules. For more strongly polar molecules it is less satisfactory, and we have seen that A may then vary with temperature. We intend to consider only the cases where A is a constant, where non-polar molecules interact solely through London forces. This value of $\phi(r)$, which will be referred to as the *LJD* 12–6 function, is then

$$\phi(r) = \frac{B}{r^{12}} - \frac{A}{r^6} .$$
(3.28)

The function $\phi(r)$ is of the shape shown in Fig. 3.1b and we can relate the quantities A and B to the molecular constants ϵ, σ and r_0 shown in this figure. From our discussion in Section 3.2 we have that when $r = \sigma$, $\phi(r) = 0$; on substituting in (3.28) this gives us

$$\frac{B}{A} = \sigma^6$$
(3.29)

Also when $r = r_0$, $d\phi/dr = 0$ which leads to

$$\frac{A}{2B} = \frac{1}{r_0{}^6} .$$
(3.30)

Thirdly, since $\phi(r_0) = -\epsilon$ we have

$$\frac{A^2}{4B} = \epsilon .$$
(3.31)

Using (3.29) and (3.31) (and eliminating r_0) we get $A = 4\epsilon\sigma^6$ and $B = 4\epsilon\sigma^{12}$, whence (3.28) reduces to

$$\phi(r) = 4\epsilon \left[\left(\frac{\sigma}{r} \right)^{12} - \left(\frac{\sigma}{r} \right)^6 \right] .$$
(3.32)

This expresses $\phi(r)$ in terms of ϵ and σ. Much work on liquids has been based on this 12–6 function, particularly for the inert gases and liquids composed of simple molecules, such as non-polar diatomic molecules.

From equations (3.29) and (3.30) $r_0{}^6 = 2\sigma^6$, that is $r_0 = 1.12\sigma$. This shows that r_0 is also a reasonable measure of the 'diameter' of the molecule. For two molecules at their normal equilibrium separation we have $r = r_0$ and $\phi(r_0) = -\epsilon$. If we double their distance apart we find, on putting $r = 2r_0$ in (3.32), that

$$\phi(2r_0) = \frac{-\epsilon}{32} ,$$

showing that the *LJD* potential falls off very rapidly with the separation r.

3.6   THE 'RIGID-SPHERE' AND OTHER SIMPLE FUNCTIONS

At this stage it is worth mentioning three simpler forms of $\phi(r)$ which have been used in work on liquids.

In the first of these, each molecule is regarded as being a non-attracting rigid sphere of diameter σ and then $\phi(r) = \infty$ for $r < \sigma$ and $\phi(r) = 0$ for $r > \sigma$. A better approximation is that shown in Fig. 3.6 where the molecules are rigid spheres of diameter σ and are acting on each other with weak attractive forces. In this case $\phi(r) = \infty$ for $r < \sigma$ and $\phi(r) = -A'/r^6$ for $r > \sigma$.

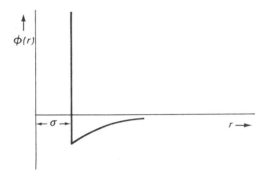

Fig. 3.6 The form of the function $\phi(r)$ for 'rigid spheres' with weak attractions.

Another potential energy function sometimes used in theoretical work is the 'square-well' function. This is defined by

$$\phi(r) = \infty, \qquad r < \dot{\sigma} \; ;$$

$$\phi(r) = -\epsilon_1, \qquad \sigma < r < \sigma_1 \; ;$$

$$\phi(r) = 0, \qquad r > \sigma_1 \; ,$$

where σ_1 is a length greater than σ and ϵ_1 is an energy. σ_1/σ is usually taken to be about 1.5.

The above three potentials are the simplest with any pretence at describing real interactions qualitatively and are much used in theoretical work. Using the 'rigid-sphere' potential some very comprehensive machine calculations using Monte Carlo methods (Wood and Parker, 1957) and molecular dynamics methods (Wainwright and Alder, 1958) have been made. These consist of explicit calculations of the equation of state of small assemblies of about 100 molecules obtained by one of two methods. The first samples their possible configurations in a truly random manner and uses the data to build up an estimate of the configurational partition function Q, the second solves the equations of motion of the individual molecules and deduces the pressure and other equilibrium properties by appropriate averaging over all the molecules.

It is also possible to develop theories of the liquid state which give good approximate results for rigid spheres; one of the most successful of these theories is that due to Percus and Yevick (1958). Again Barker and Henderson (1968) have used the rigid-sphere potential as the basis of a perturbational calculation of the equation of state of a real liquid.

We next ask: how is the definition of Q, given in equation (3.6), extended to these discontinuous potentials? For the rigid spheres one conventionally takes Φ as infinite, that is $\exp(-\Phi/kT)$ as zero, for all configurations of the sphere centres which imply two or more spheres overlapping. Put more mathematically, the integral in (3.6) reduces to an estimate of the ($3N$-dimensional) volume of configuration space that is accessible to the centres of the N spheres.

For the square-well model, configurations corresponding to overlap are excluded as before. However configurations in which a given pair of spheres have centres between σ and σ_1 apart are weighted more heavily than are those for which this distance apart is greater than σ_1. Speaking mathematically, the multiple integral (3.6) must be defined in an appropriate 'piecewise' sense for these potentials.

3.7   THE LAW OF CORRESPONDING STATES

It is convenient to emphasise now that the work we have been discussing in this chapter is based on certain assumptions which for the sake of clarity we will state here again. They are:
(1) The methods of classical statistical mechanics may be used.
(2) Interactions between different pairs of molecules are additive, that is equation (3.4) holds.
(3) The interaction energy of each pair of molecules depends only on the distance between their centres.
(4) There are no internal degrees of freedom.

The function $\phi(r)$ in equation (3.32) is a product of an energy ϵ and a function of the dimensionless quantity r/σ. Let us then assume that $\phi(r)$ can in general be represented by some universal function $f(r/\sigma)$ multiplied by ϵ, that is

$$\phi(r) = \epsilon f(r/\sigma) . \tag{3.33}$$

This $\phi(r)$ depends on r only. We now examine the consequences of assuming that the function f is the same for two liquids. We can still choose ϵ and σ for each liquid. The 12–6 potential of equation (3.32) merely represents a special case of the behaviour of $\phi(r)$ given by equation (3.33).

From equations (3.7) and (3.9) the equation of state is given by

$$P = kT\left(\frac{\partial lnZ}{\partial V}\right)_T .$$

The only term on the right-hand side of equation (3.5) which depends on the volume is Q. Thus differentiating lnZ with respect to V leads to $\partial lnQ/\partial V$ only; that is

$$P = kT\left(\frac{\partial lnQ}{\partial V}\right)_T$$

$$= \frac{kT}{Q}\left(\frac{\partial Q}{\partial V}\right)_T . \qquad (3.34)$$

Dimensionally, we have

$$[P] = \frac{[Energy]}{[Length]^3} \quad , \quad \left[\frac{V}{N}\right] = [length]^3 \quad , \quad [T] = \frac{[Energy]}{[k]}$$

where there are N molecules in a volume V.

This suggests that we could use ϵ/σ^3, σ^3 and ϵ/k respectively as the units (so-called 'reduced' or 'molecular' units) in which to measure P, V/N and T. Hence if we write

$$P^* = \sigma^3 P/\epsilon$$

$$V^* = V/N\sigma^3$$

$$T^* = kT/\epsilon$$

and
$$r_{ij}^* = r_{ij}/\sigma$$

the quantities P^*, V^*, T^* and r_{ij}^* are dimensionless and are pure numbers. They are called 'reduced properties' or 'reduced variables' in molecular units and result from making the original variables dimensionless by expressing them in units which are combinations of the molecular constants σ and ϵ. In Table 3.2 reduced units for various properties are shown:

Table 3.2

Quantity	Unit
Intermolecular distance, r_{ij}	σ
Temperature, T	ϵ/k
Pressure, P	ϵ/σ^3
Volume, V	$N\sigma^3$
Density, $d = 1/V$	$1/N\sigma^3$
Energy, U	$N\epsilon$
Entropy, S	Nk

If we now substitute $P = \epsilon P^*/\sigma^3$, $kT = \epsilon T^*$, $V = N\sigma^3 \, V^*$ into equation (3.34) we readily obtain

$$P^* = \frac{T^*}{NQ}\left(\frac{\partial Q}{\partial V^*}\right)_{T^*} \tag{3.34a}$$

in terms of the reduced quantities P^*, T^* and V^*. We still have to consider Q in terms of reduced units. Now, from (3.6),

$$Q = \frac{1}{N!}\int \cdots \int \exp\left[-\Sigma\phi(r_{ij})/kT\right] dx_1 \ldots dz_N \tag{3.35}$$

Also
$$\frac{\Sigma\phi(r_{ij})}{kT} = \frac{\Sigma\epsilon f(r_{ij}/\sigma)}{\epsilon T^*} = \frac{\Sigma f(r_{ij}^*)}{T^*}$$

and for every length such as dx_1 we may write σdx_1^*. Thus on substitution into (3.35) we have

$$Q = \frac{\sigma^{3N}}{N!}\int \cdots \int \exp\left[-\Sigma f(r_{ij}^*)/T^*\right] dx_1^* \ldots dz_N^*$$

$$= \sigma^{3N}Q^* \text{ (say)}.$$

Hence on substituting this value of Q in both the numerator and denominator of equation (3.34a) we have

$$P^* = \frac{T^*}{NQ^*}\left(\frac{\partial Q^*}{\partial V^*}\right)_{T^*}$$

The right-hand side of this equation does not depend on ϵ and σ but only on reduced quantities. Thus we may write the reduced pressure as a universal function of the reduced volume and reduced temperature, that is

$$P^* = P^*(V^*, T^*) \tag{3.36}$$

This relation is true for all substances which are described by the common function f and is independent of the numerical values of ϵ and σ. The fact that universal functions such as $P^*(V^*, T^*)$ exist for a group of substances having the same common function f in their potential function is called **the law of corresponding states**. (This law is a *rigorous* deduction from the statistical mechanical expressions for the partition function and configuration integral.) The thermodynamic properties of all the substances in the group are then identical when expressed in terms of the same reduced units (e.g. those in Table 3.2).

In theoretical work it is more convenient to present results in terms of reduced variables, but experimental results are usually obtained in terms of the actual state variables. Thus when the molecular constants ϵ, σ etc. for the substance concerned are known, the experimental results can be reduced and a comparison between theory and experiment made. It will be our practice to show all reduced variables by means of an asterisk; for example, reduced entropy is always shown as S^*. Further, it is often convenient to express a result as an 'excess' thermodynamic function. This is the difference between the actual value of that particular function for a real fluid and the value it would have if the substance existed as a perfect gas at the same temperature and density. It is a difference which arises, of course, because of the intermolecular forces.

3.8   THE VALUES OF ϵ AND σ IN THE *LJD* POTENTIAL

Before we can make a comparison between the experimental values of various thermodynamic properties and the corresponding theoretical values based on the *LJD* 12–6 potential the parameters ϵ and σ in (3.32) must first be determined. These two quantities are usually found by fitting the second virial coefficient calculated from (3.32) to the experimental value of the second virial coefficient of that particular substance when it is in the gaseous phase (see Section 4.2).

Values of these parameters for some simple liquids are shown in Table 3.3.

Table 3.3 Parameters for the *LJD* 12–6 potential.

Substance	$\epsilon/k_{(^\circ K)}$	$\sigma_{(\text{Å})}$
Xe*	225.3	4.070
Kr*	166.7	3.679
Ar*	119.5	3.405
Ne[+]	35.60	2.749
H_2^+	37.00	2.928
He[+]	10.22	2.556

* Walley and Schneider (1955)
[+] Hirschfelder *et al* (1954)

A second way of obtaining ϵ and σ is to consider the crystalline solid phase at absolute zero. To illustrate the method let us concentrate our attention on a crystal of N atoms of solid argon at $0°K$; this is a face-centred cubic lattice for which $z = 12$. We assume that each pair of atoms interacts according to the 12–6 law and, by means of a lattice summation, the zero-point potential energy Φ_0 can be calculated in terms of N, ϵ, σ and a (the nearest neighbour distance). To this classical energy Φ_0 we must add the zero-point energy of the crystal, K, whose value can be obtained from the quantum theory of specific heats. So at $0°K$ the total internal energy of the crystal is $U_0 = \Phi_0 + K$. This total energy is also equal to the energy required to separate the N atoms so that they are at infinite distances from each other; in other words U_0 is equal to L_s, the zero-point latent heat of sublimation. Thus we can write

$$U_0 = \Phi_0 + K = L_s. \qquad (3.37)$$

Also, at the zero-point the value a_0 of a must be such that U_0 is a minimum. This condition, taken together with equation (3.37), enables us to determine ϵ and σ. For further details see *The Liquid State*, Chapter 4, by J. A. Pryde.

The ultimate test of any theory is how well it agrees with experiment. The law of corresponding states works well in practice because it results from the natural assumption that the interaction function of a given type of molecule can be related to that of others of similar shape and structure by simple scaling laws. One consequence of the law is that the reduced parameters of the triple and critical points should be the same for all substances, and Table 3.4 shows that this prediction is fulfilled very well for the heavier inert gases.

Table 3.4

Substance	T^*_{trip}	$P^*_{trip} \times 10^3$	V^*_{trip}	T^*_{crit}	P^*_{crit}	V^*_{crit}
Ne	0.690	1.882	1.292	1.247	0.1150	3.333
Ar	0.699	1.646	1.182	1.258	0.1162	3.164
Kr	0.698	1.556	1.161	1.260	0.1167	3.163
Xe	0.702	1.570	1.162	1.260	0.1134	3.095

Other predictions for these inert gases are also verified very accurately; for further details the reader is referred to *Statistical Theory of Liquids,* Chapter 1, by Fisher.

What we can conclude from the fact that the equation of state of liquid argon, for example, is reasonably reproduced if an *LJD* interaction function (equation 3.32) is used in the calculation of Q is less clear. At first sight this is very encouraging, but recent evidence suggests that there may be some cancellation of errors. The fact that the interaction is not entirely two-body seems at liquid densities just about to balance the effect of discrepancies between the *LJD* function and the interaction between two real molecules! Such situations are typical of liquid and solid state physics.

3.9   THE PAIR POTENTIAL FUNCTION FOR LIQUID METALS

So far, our remarks in this Chapter have been concerned with insulating (that is, non-metallic) liquids which do not conduct electricity. In recent years however, conducting liquids in the form of liquid, or molten, metals have been widely studied (see, for example, March (1968). Since liquid metals have certain fluid properties in common with non-metallic liquids we will digress slightly and devote this Section to a brief discussion of $\phi(r)$ for liquid metals.

The high conductivity of a metal, whether solid or liquid, arises from the motion of the 'free' electrons — the conduction electrons. These electrons have been released from their parent atoms which, thus deprived of some of their electrons, are now positive ions. Hence a liquid metal consists of positive ions and conduction electrons, and may be regarded as two fluids thoroughly mixed throughout. One of these fluids consist of ions, the other is a gas of conduction electrons. To deal with this situation using the method of pair interactions the concept of a 'pseudo-atom' has been introduced (see Egelstaff, 1967). According to this picture the electron gas is regarded as a charge cloud pervading the whole volume of the liquid metal. Every ion is surrounded by its own local part of this cloud which produces a screening effect around the ion. Each ion, together with its negative screening cloud, comprises a 'pseudo-atom'.

To work out the interionic potential $\phi(r)$ for this assembly requires quantum mechanics and the treatment is beyond the scope of this book. We shall merely state that the result shows that $\phi(r)$ so calculated oscillates with the distance r. A rough comparison between the $\phi(r)$ for the interatomic potential between the atoms of liquid argon and that for the interionic potentials between the ions in liquid lead is shown in Fig. 3.7. For further details the reader is referred to an article by Cusack (1970).

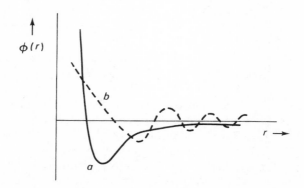

Fig. 3.7 Variation of $\phi(r)$ with r for argon atoms and for lead ions in liquid lead. (Curve a: Argon; Curve b: Lead)

Liquid sodium may be taken as the 'classic' example of the type of liquid metal that we are describing, and recent work by Greenfield *et al* (1971) and by Howells and Enderby (1972) is of considerable interest. Greenfield and his colleagues obtained very accurate X-ray structure data for liquid sodium. Howells and Enderby then calculated the value of the effective pair potential $\phi(r)$ for this liquid metal by combining the X-ray data with various integral equations and also obtained oscillations with distances.

Quite a number of theoretical workers have suggested that these 'damped oscillations' with r in the effective interaction energy are to be expected as a consequence of the fact that the conduction electrons obey Fermi statistics.

REFERENCES

Books

Born, M., (1969), *Atomic Physics, 8th Edition*, Appendix XL, Blackie, London.
Egelstaff, P. A., (1967), *An Introduction to the Liquid State*, Ch. 4, Academic Press, London and New York.

Fisher, I. Z., (1964), *Statistical Theory of Liquids*, The University of Chicago Press, Chicago and London.

Flowers, B. H. and Mendoza, E., (1970), *Properties of Matter*, Ch. 3, John Wiley and Sons Ltd.

Hirschfelder, J. O., Curtiss, C. F. and Bird, R. B., (1954), *Molecular Theory of Gases and Liquids*, Wiley, New York.

March, N. H., (1968), *Liquid Metals*, Pergamon Press.

Pryde, J. A., (1966), *The Liquid State,* Hutchinson University Library, London.

Rowlinson, J. S., (1959), *Liquids and Liquid Mixtures*, Butterworths Scientific Publications, London.

Papers

Barker, J. A. and Henderson, D., 1968, *Journ. Chemical Education,* **45**, 2.

Cusack, N. E., 1970, *The Physics of Liquid Metals,* an article in *Liquids,* Taylor and Francis Ltd., London.

Greenfield, A. J., Wellendorf, J. and Wiser, N., 1971, *Phys. Rev.* **A4**, 1607.

Howells, W. S. and Enderby, J. E., 1972, *J. Phys. C.,* **5**, 1277.

London, F., 1930, *Z. Phys. Chem. (B)* **11**, 222.

London, F. 1937, *Trans. Farad. Soc.,* **33**, 8.

Percus, J. K. and Yevick, G. J., 1958, *Phys. Rev.,* **110**, 1.

Tabor, D., 1964, *Contemp. Phys.,* **6**, 112.

Wainwright, T. W. and Alder, B. J., 1958, *Nuovo Cim.,* **9**, Suppl. 1, 116.

Walley, E. and Schneider, W. G., 1955, *J. Chem. Phys.,* **23**, 1644.

Wood, W. W. and Parker, F. R., 1957, *J. Chem. Phys.,* **27**, 720.

Chapter 4
VARIOUS APPROACHES TO THE
STATISTICAL MECHANICAL PROBLEM

In Chapter 3 we introduced the partition function and discussed the possible forms of intermolecular interaction (which we must know to calculate this partition function). The fundamental result of statistical mechanics means that all the equilibrium properties of an assembly can be calculated from the partition function. Because of this we ask how accurately we can calculate the partition function if molecules are supposed to interact only in pairs and the interaction energy is a given function of distance.

In principle we could calculate the partition function associated with any given interaction function with sufficient accuracy if we had access to a big enough computer. At present such calculations are possible only for reasonably simple interaction functions and for 'model' assemblies of about 100 molecules. We must therefore examine two considerations, first what approximate methods are possible and secondly whether any general deductions can be made about the liquid state without resorting to drastic approximations.

The early development of liquid state theory was based very largely on heuristic models, and agreement of their consequences with experiment was only moderate. They have been largely superseded by approaches based on integral equations and by large scale computer simulations, but this is not to say that they no longer have any value. In fact these models give real physical insight into the processes occurring in a liquid, so that their interest is considerably more than historical. We describe some of these models and summarise some exact results and modern analytic approaches.

4.1 EXACT RESULTS

4.1.1 Corresponding States

In Chapter 3.7 we have already seen that if two substances have intermolecular interactions described by $\epsilon_1 f(r/\sigma_1)$ and $\epsilon_2 f(r/\sigma_2)$, where the function f is the same for both, the partition functions (and therefore the other equilibrium properties) of the two substances become *identical* functions if the density and temperature are scaled by appropriate factors. This is a rigorous consequence of the form for the configurational partition function given by equation (3.6). In fact similar scaling laws would always hold for an interaction function that depended on no more than two parameters, and we shall meet a specific example below.

4.1.2 The Mayers' Virial Expansion

The reader must turn to the Mayers' book *Statistical Mechanics* for a full account of their work relating the properties of imperfect gases to the intermolecular interaction function. Their method can, in principle, be extended to describe the liquid state itself. Their speculations about the analytic nature of the critical region are not now accepted, but this in no way affects the basic soundness of their approach to the imperfect gas and liquid problems. We give an account of some of the key moves in this approach, referring the reader to their book, or to other advanced treatises on statistical mechanics, for the mathematical details. Equation (3.5) demonstrates that the partition function of a classical assembly is known if we can calculate the configuration integral Q (N, V, T). This is simply the Boltzmann factor associated with the intermolecular interaction energy integrated over all possible positions of the N molecules in an enclosure of volume V. Thus

$$Q(N, V, T) = \frac{1}{N!} \underbrace{\int \ ... \ \int}_{\leftarrow \ 3N \ \rightarrow} \exp\left[-\sum_{i<j} \phi(r_{ij})/kT\right] dx_1 \ ...dz_N \qquad (4.1)$$

(c.f. equation (3.6)), where we have assumed that the interaction between molecules i and j is independent of the positions of all the other molecules and that it depends only on their mutual distance r_{ij}. Now we know that the interaction energy $\phi(r_{ij})$ becomes very large and positive when r_{ij} is small and that, after going through a negative minimum as r increases, it rapidly approaches zero from below when r_{ij} is larger than a few molecular diameters (see Fig. 3.1b).

We define the Mayer function $f(r_{ij})$ or f_{ij} by the equation $\exp[-\phi(r_{ij})/kT] = 1 + f_{ij}$. Because of the behaviour of the function ϕ, f is -1 for r_{ij} small compared with the gas-kinetic diameter σ and approaches zero for r_{ij} large. We express the integrand of (4.1) in terms of the f-functions and group the terms according to the numbers of f-functions they contain. Thus we have

$$\text{Integrand} = 1 + \sum_{i \neq j} f_{ij} + f_{12}f_{23} + ... + f_{12}f_{34} + ... \qquad (4.2)$$

This has to be integrated over all positions of all the molecules. The first term in (4.2) is what we should get for a perfect gas; when integrated it gives simply V^N. Any one of the terms f_{ij} when integrated would lead to

$$V^{N-2} \int f_{ij} \; \mathrm{d}x_i \; \mathrm{d}y_i \; \mathrm{d}z_i \; \mathrm{d}x_j \; \mathrm{d}y_j \; \mathrm{d}z_j$$

(6)

and, because the function f_{ij} is the same for all pairs of molecules, all these $N(N-1)/2$ terms are numerically the same. The Mayers prove that this integral is proportional to the second virial coefficient in the equation of state of the gas.

The various terms in (4.2) can be regarded as describing in successive approximation the departure of the assembly from perfect gas behaviour. The function f_{ij} is only appreciable when r_{ij} is not too large compared with σ, so that the contribution of (say) the f_{12} term to the integral (4.1) is of the order of σ^3/V times that of the leading term. Similarly, the contribution of a term like $f_{12}f_{23}$ when integrated will be of the order of σ^6/V^2 times the leading term, because *both* molecules 1 and 3 have to be near to 2 for the product $f_{12}f_{23}$ to be appreciable. Analogous results hold for terms containing more f-factors.

The full argument can be made mathematically rigorous and shows indeed that the departure of an assembly from perfect gas behaviour can be expressed in the form of a series, the higher terms of which correspond in succession to integrals of products of f-factors which are only appreciable when 2, 3, 4 or more molecules are near to one another. Consequently, these higher terms are known as **cluster integrals** of order 2, 3, 4 and are written $b_2, b_3, b_4 \ldots$ etc. As we might expect, the higher order cluster integrals become relatively more and more important in determining the equation of state as the density is increased. Specifically we find

$$\frac{P}{kT} = \sum_{\varrho} b_{\varrho} \, z^{\varrho}$$

(4.3)

where z is a parameter known as the **fugacity**. This is related to the density by the equation

$$\frac{N}{V} = \sum_{\varrho} \varrho b_{\varrho} \, z^{\varrho}$$

(4.4)

where b_1 is defined to be unity, and the higher b's are defined as above. For very low densities, both series (4.3) and (4.4) reduce to their first terms and we have the perfect gas law. For somewhat higher densities, we have in turn the corrections to the perfect gas law associated with the interactions of groups of 2, 3, 4 and so on molecules with one another. Mathematically, we obtain a virial series of the type (4.5) by eliminating the variable z between (4.3) and (4.4).

In fact the Mayers rigorously show that the result of eliminating the variable z between equations (4.3) and (4.4) leads to an expansion of pressure in ascending powers of the density N/V, and that this expansion is certainly valid if series (4.3) and (4.4) are convergent. This is certainly true at low enough densities for a wide range of possible interaction functions. We conclude that, at low enough densities, the departure of a gas from perfection should be described by the equation

$$\frac{P}{kT} = \frac{N}{V} + B\left[\frac{N}{V}\right]^2 + C\left[\frac{N}{V}\right]^3 + D\left[\frac{N}{V}\right]^4 \ \dots \dots \tag{4.5}$$

where B is related to the Mayer function f by

$$B = -2\pi N \int_0^\infty r^2 \ f(r) \ \mathrm{d}r \tag{4.6}$$

and the higher virial coefficients C, D etc. are likewise expressible as multiple integrals of products of Mayer functions. For example, C is proportional to the quantity $f_{12} f_{23} f_{13}$ integrated over all possible positions of molecules 1, 2 and 3.

For the purpose of comparing theory with experimental equation of state data we need only the first few virial coefficients. However, knowledge of the behaviour of the higher order ones is important both for determining the range of validity of equation (4.5) and for the discussion of phase transitions as we shall see in the next Chapter. Unfortunately, very little precise information about the higher order virial coefficients is available.

4.1.3 The model of Longuet-Higgins and Widom

However, there is one very important case in which precise information about higher virial coefficients *is* available to all orders. We suppose that the attractive part of the interaction between two molecules is very weak, but of very long range. Longuet-Higgins and Widom make a model of argon by combining this assumption about the attractive forces with the assumption that, *qua* repulsive forces, the argon atoms behave like rigid spheres. It can then be shown rigorously that the equation of state of this model is

$$P = T \phi\left(\frac{V}{V_0}\right) - \frac{a}{V^2} \tag{4.7}$$

where the term $T \phi\left(\dfrac{V}{V_0}\right)$ is the pressure of a gas of non-attracting rigid spheres, and the term a/V^2 represents the effect of long-range attractions. Admittedly the function $\phi(V/V_0)$ is not yet known analytically, but the first seven virial coefficients are known. Moreover the behaviour when the assembly is nearly close packed is known from the computer calculations of Alder *et al*. Hence ϕ can be regarded as an accurately known function, since it is known for both large and small values of V/V_0.

In fact Longuet-Higgins and Widom find that equation (4.7) gives an extremely good account of the properties of argon, and of the other rare gases also with appropriate scaling.

4.2   THE VAN DER WAALS MODEL

Evidently this differs from equation (4.7) simply by the replacement of $T \phi(V/V_0)$ by the less accurate expression $NkT/(V-b)$. In the region of the critical point this makes very little difference. The van der Waals equation involves just two parameters, a and b, and therefore should satisfy the principle of corresponding states. We now verify that this is so. Although equation (4.7) only involves the two parameters V_0 and a and thus also satisfies the principle, the calculations can only be done numerically because the function ϕ is only known numerically. For the van der Waals model we saw in Chapter 2.2 that the critical isothermal has a critical point C (Fig. 2.4). At C this isothermal has a horizontal tangent and C is a point of inflexion as well. These facts enable us to obtain the critical pressure, volume and temperature corresponding to the point C. The results are

$$P_c = \frac{a}{27b^2}, \quad V_c = 3b, \quad T_c = \frac{8a}{27bR}. \tag{4.8}$$

The values of a and b can be found by substituting observed values of P_c and T_c into these equations and the predicted value of $3b$ for V_c then compared with the observed value for V_c; typical observed values for V_c are 2.34 b for argon and 2.44 b for hydrogen.

Equations (4.8) also predict that the dimensionless quantity $RT_c/P_c V_c$ has the values $8/3 = 2.67$; the observed values of this quantity are all around 3.5. All these considerations show that van der Waals's equation is not very closely obeyed by a gas near the critical point.

If now we divide the values of P, V and T by the corresponding critical values we obtain three dimensionless quantities, P', V', and T', called **reduced variables**. Thus

$$\frac{P}{P_c} = P', \text{the reduced pressure}$$

$$\frac{V}{V_c} = V', \text{the reduced volume} \tag{4.9}$$

$$\frac{T}{T_c} = T', \text{ the reduced temperature.}$$

Using (4.9) to substitute for P, V and T in (2.5) we obtain

$$\left(P_c P' + \frac{a}{V_c^2 V'^2}\right)(V_c V' - b) = RT_c T'$$

and further, using (4.8) we have

$$\left(\frac{P'a}{27b^2} + \frac{a}{9b^2 V'^2}\right)(3bV' - b) = \frac{8aT'}{27b}.$$

This reduces to
$$\left(P' + \frac{3}{V'^2}\right)(3V' - 1) = 8T'. \tag{4.10}$$

This is a **reduced form of van der Waals's equation**. This equation should be true for *all* gases since it does not contain the constants a and b for any particular gas. Different gases having the same values of P', V' and T' are said to be in **corresponding states**.

If, using equation (4.10), we plot a number of (P', V') curves for various fixed values of T', these 'reduced' isothermals will in theory be the same for all substances. In actual practice (that is, by using the *experimental* values of P', V', and T') the reduced isothermals for different substances do follow each other very closely. This close agreement is such that it is justifiable to state that 'if two or more substances have the same values for any two of the reduced variables P', V' and T' they therefore have approximately identical values for the third'. This is a particular case of the **Law of Corresponding States** which we found in Chapter 3.7 to follow as a consequence of a certain assumption about the interaction energies.

The quantities P', V' and T' are just one example of the form in which reduced variables can be expressed; other reduced variables based on molecular properties were also discussed in Chapter 3.7.

4.3   COMPARISON OF THE VIRIAL EQUATION WITH EXPERIMENT

We must now consider other equations for representing the behaviour of imperfect gases. We have seen from Mayer's work that if a gas is at fairly low density where its behaviour is not very different from that of a perfect gas then its equation of state may be written as

$$\frac{PV}{RT} = 1 + \frac{NB}{V} + \frac{N^2C}{V^2} + \frac{N^3D}{V^3} + \ ...... \tag{4.11}$$

or, equivalently, as

$$\frac{PV}{RT} = 1 + B'P + C'P^2 + D'P^3 + \ ..... \tag{4.12}$$

These equations are called virial equations of state. B is called the second virial coefficient, C the third virial coefficient, and so on. The first virial coefficient is simply unity, so that for a perfect gas all the virial coefficients apart from the first are zero. Thus the second, third etc. coefficients are a measure of how much the behaviour of an imperfect gas departs from that of a perfect gas and they are all functions of temperature, that is $B = B(T)$ etc.

We have seen that a perfect gas obeys Boyle's law, $PV = $ const. (for any given constant temperature). Real gases such as air, oxygen, nitrogen and hydrogen obey the law very closely indeed at ordinary pressures and temperatures, but at pressures of the order of thousands of atmospheres there are marked deviations from the law. The results show that, for a given mass of gas at constant temperature, the value of PV is not a constant but is some function of pressure. It is also found that there is a certain temperature T_B for each gas at which the product PV is a constant and is independent of P; in other words at $T = T_B$ Boyle's law holds. Now consider these facts in conjunction with equations (4.11).

For very small pressures PV has the same value of RT for all gases, that is, $PV = $ const. for constant temperature. The second virial coefficient B is the most important when considering the behaviour of individual gases, since C, D etc. are so small that the terms involving them can be neglected under such conditions. At low temperatures B is negative, at higher temperatures it is positive and at some intermediate temperature it is zero. At this intermediate temperature, **which we therefore identify with the Boyle temperature T_B,** we have, from (4.11).

$$\frac{PV}{RT_B} = 1 \text{ or } PV = \text{ const.}$$

since we are neglecting terms in C, D etc. That is, Boyle's law is fairly well obeyed at the temperature T_B for which $B = 0$ provided that P is not so great (or V so small) that the terms involving the third, fourth etc. coefficients make significant contributions.

The second virial coefficient B is of great importance also for the following reason. Equation (4.6) tells us that Mayer's theory of imperfect gases shows that B can be expressed as

$$B = -2\pi N \int_0^\infty [\exp(-\phi(r)/kT) - 1] r^2 \, dr$$

for a gas of N molecules. Thus comparison between the theoretical and experimental values of this coefficient can yield information about the potential $\phi(r)$ between two molecules. In particular, if $\phi(r)$ is given by equation (3.32) this comparison can lead to a determination of the quantities ϵ and σ in (3.32), as we saw in Chapter 3.8.

Equations of types (4.11) and (4.12) were first given empirically by Kammerlingh-Onnes at the beginning of the century, long before the theoretical justification of them found by the Mayers. In what follows the expansion (4.11) will be taken as far as the fourth term only and we will, following de Boer and Uhlenbeck (1964), write it as

$$\frac{PV}{NkT} = 1 + Bd + Cd^2 + Dd^3 \tag{4.13}$$

where $d = N/V$ and N is the number of molecules in the volume V under consideration.

As already indicated, the calculation of B, C and D involves a knowledge of the function f_{ij}. This depends in turn on the choice of the function $\phi(r_{ij})$, which we will take to be the *LJD* 12–6 potential (equation (3.32)). Using this 12–6 potential the values of B, C and D have been calculated. (Accurate numerical results for B and C are available over a large temperature range while much less information about D is available). Using these theoretical values of B, C and D one can then obtain the equation of state and other properties of a dense gas; these theoretical values can then be compared with the experimental values.

Such a comparison has been given by Levelt and Cohen (1964) in the two sets of graphs shown in Fig. 4.1. In Fig. 4.1a the value of the excess energy U_e*, in reduced units, is plotted against the reduced density $d* = N\sigma^3/V$ for the gases argon and xenon. In Fig. 4.1b the reduced excess entropy S_e* is plotted against $d*$ for the same two gases. (The 'excess' value of any thermodynamic function is defined as the difference between the value of that function for a substance and the value it would have if the substance existed as a monatomic perfect gas at the same density and temperature). The full curves are the experimental results while the circled points represent the theoretical results from the four-term virial expansion in (4.13). In each case results for two values of the reduced temperatures are shown, namely $T* = 3.5$ (near the

Boyle temperature) and $T^* = 1.35$ (just above the critical temperature). These graphs show that up to reduced densities of 0.4 or so there is very good agreement between theory and experiment.

The weakness in this comparison between theory and experiment lies in the fact that only four terms in the virial expansion were taken. The last term, Dd^3, involves the interactions in a group of four molecules. At the high densities in which we are interested the interactions between groups of molecules containing more than four molecules are obviously important. Hence the good agreement between experiment and the results based on a four-term virial expansion could conceivably be accidental. However there is some evidence that, for reasonable temperatures and physically likely interaction functions, the higher virial coefficients may be numerically small and of mixed signs. If so, their combined effect may well be negligible. It must be emphasised that in this comparison between theory and experiment we have actually considered a dense gas rather than a true liquid; all our comments refer to the case of supercritical temperatures. For further details the reader is referred to *Studies in Statistical Mechanics*, Volume II, Chapter 2, Part B (edited by de Boer and Uhlenbeck).

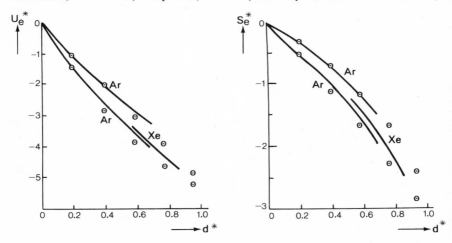

Fig. 4.1 (a) Excess energy as a function of the reduced density. (b) Excess entropy as a function of the reduced density. Full curves: experimental excess energy for argon and xenon. Circles: Theoretical values using four virial coefficients. Upper curves and circles $T^* = 3.5$. Lower curves and circles $T^* = 1.35$ (After Levelt and Cohen).

4.4   THE CELL MODEL OF A LIQUID

The cell model for a liquid was introduced by Lennard-Jones and Devonshire (1937, 1938). It is an attempt to develop a one-particle liquid model analogous to the Einstein model for a solid.

Consider a liquid of volume V consisting of N spherical molecules considered first as **rigid spheres**. Then any one of these molecules is confined to a finite volume, or 'cell', formed by the molecules which are its nearest neighbours (Fig. 4.2). The following assumptions are then made:

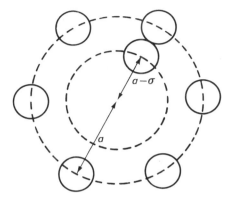

Fig. 4.2 Central plane section of a typical cell.

(1) The total volume is made up of identical cells, each of which contains one 'central' molecule only.
(2) The centres of the cells form a close-packed lattice for which $z = 12$.
(3) The molecules are regarded as moving independently in their cells.
We shall not discuss these assumptions until later, except to state that assumption (2) is to be understood as implying that the immediate environment around a chosen molecule is not unlike that in a solid. However the order is local only and not long-range.

The configurational partition function corresponding to this assembly of rigid spheres can be shown to be

$$Q(N, V, T) = v_f^N \tag{4.14}$$

where the 'free volume' v_f is the volume available to the centre of each molecule in its cell when its neighbours are at the centres of their cells. Clearly $v_f < V/N$, the volume per molecule, since the molecules in the liquid are very closely packed.

An estimate of v_f can be obtained as follows. Choose a given 'central' molecule and assume that the centres of its twelve nearest neighbours lie on a sphere of radius a about the cell centre; this is the outer sphere shown in Fig. 4.2. Then the *minimum* distance the central molecule can move from the cell centre to collision with a neighbour is $a-\sigma$, where σ is the diameter of a molecule. Thus the minimum value of v_f is the inner spherical volume in Fig. 4.2 that is

$$v_f = \frac{4\pi}{3}(a-\sigma)^3 . \tag{4.15}$$

This value of v_f is, as already explained, an underestimate. If we assume that the cell centres lie on a regular close-packed lattice it follows from the geometry that

$$a = \left[\sqrt{2}\frac{V}{N}\right]^{1/3} .$$

When nearest neighbours throughout the liquid are in contact, the volume of the liquid has its minimum value V_0. This occurs when $a = \sigma$, so that

$$\sigma = \left[\sqrt{2}\frac{V_0}{N}\right]^{1/3} .$$

Substituting in (4.15) we get

$$v_f = \frac{4\pi\sqrt{2}}{3N}\left[V^{1/3} - V_0^{1/3}\right]^3 .$$

In this way v_f has been related to V/N, the volume per molecule. Using this value of v_f, Q can be determined from (4.14), and hence the various thermodynamic properties derived.

Lennard-Jones and Devonshire also considered a more sophisticated cell model assuming 12–6 interactions between the molecules. To obtain v_f in this case they introduced the idea of 'smearing', which we will now discuss briefly. For the close-packed lattice there are twelve nearest neighbours and each cell is a dodecahedron formed by the planes with perpendicularly bisect the lines joining adjacent cell centres. In the smearing approximation this dodecahedron is replaced by a sphere, and Lennard-Jones and Devonshire suggested that the twelve neighbouring molecules could be 'smeared' with uniform probability distribution over the surface of this sphere. We shall not give all the details of the method but merely state that v_f and hence $Q(N, V, T)$ are thus obtained. The various thermodynamic quantities can then again be derived.

In the comparison which follows we shall concentrate on argon. The 12–6 formula is a pretty close approximation to the actual interaction between argon atoms if, following Michels *et al.* (1949), we put $\epsilon = 119.8k$ and $\sigma = 3.405\text{Å}$ (see Equation 3.32). For argon the triple point is $83.82°K$, giving $kT/\epsilon = 0.700$. At this triple point the vapour pressure is very small and has a negligible effect on the properties of the condensed phases. Hence we are not far wrong if we compare the theoretical *LJD* properties of the *zero-pressure* condensed phase for

$kT/\epsilon = 0.7$ with those of solid and liquid argon at the triple point. This comparison is made in Table 4.1 below.

Table 4.1

Property	LJD Cell theory based on 12-6 potential	Experimental Values	
		Solid argon	Liquid argon
Reduced Volume, V/V_0	1.037	1.035	1.186
Reduced excess energy, $U_e*/N\epsilon$	−7.32	−7.14	−5.96
Reduced excess entropy, S_e*/Nk	−5.51	−5.33	−3.64

From this table we see that the *LJD* theoretical value for the volume is almost the same as the experimental solid value and about 15 per cent lower than the experimental liquid value. Furthermore the *LJD* values for the energy and entropy are much closer to the solid than to the liquid experimental values. This is also true for the vapour pressures. We must therefore conclude that the condensed phase described by the *LJD* theory is more like a solid than a liquid.

One of the assumptions made in the *LJD* theory is that each molecule is surrounded by twelve nearest neighbours. If the liquid expands the density decreases, and it is more likely that the average number of nearest neighbours z will be less than twelve. Indeed, this is borne out by the results of X-ray and neutron diffraction work. In fact de Boer (1952) suggested that the *LJD* theory would give results more in agreement with the experimental liquid values if lower values of z were taken, and this is the case. Another way of achieving the same result is to suppose that some 'holes', or empty cells, exist in the liquid; this is clearly equivalent to saying that z is reduced. Such a 'hole' model is discussed in the next section.

The existence of a lower number of nearest neighbours in a liquid as compared with a solid has been shown in an elegant way by Walton and Woodruff (1969). They built a liquid-simulator in which they studied a two-dimensional 'monatomic liquid' consisting of oil-covered ball-bearings on a randomly shaken tray. The 'attractive force' between the molecules was provided by the oil coating and the 'thermal energy' by vibrating the bearings on the tray. When the oil-covered ball-bearings were first placed on the *stationary* tray they formed a roughly regular close-packed two dimensional lattice, that is a 'solid' with $z = 6$. As the vibration (the 'temperature') was increased no change occurred until a definite 'temperature' was reached, when the structure melted into a 'liquid'. A typical picture of this simulated liquid is shown in Fig. 4.3, a still from a cine film of the events as they took place. Such individual frames of the 'liquid' show how each ball is surrounded, on average, by five nearest neighbours and not six as in the close-packed two-dimensional solid. Further the behaviour of any one ball shows that for most of its time it is confined within its cell of nearest neighbours. Figure 4.4 shows the distribution function for this model, and it is qualitatively similar to that of a real liquid.

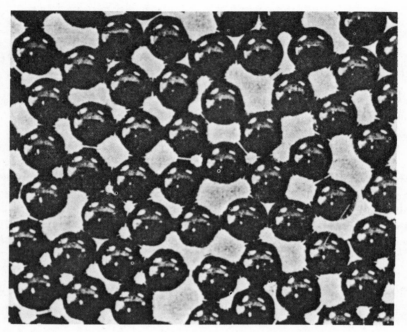

Fig. 4.3 Instantaneous snapshot of a 'two-dimensional liquid', after Walton and Woodruff.

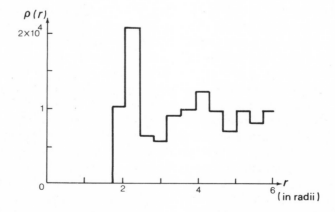

Fig. 4.4 Radial variation of density, after Walton and Woodruff.

Finally, it must be emphasised that in this section we have given only an outline of the cell model for a liquid and that there are many modifications and refinements that have not been mentioned. A much more detailed account of the model may be found in *Lattice Theories of the Liquid State* by J. A. Barker (1963).

4.5   THE HOLE MODEL

In this case the liquid is pictured as consisting of a lattice of cells. A cell can then either be empty, in other words be a 'hole', or can contain no more than one molecule. Let us suppose that there are a total of N molecules and N_0 holes occupying a total of $(N + N_0)$ lattice cells. Properties like thermal expansion and the decrease of density with rising temperature are then explained as being due to an increase in N_0 as the temperature increases.

The various attempts made using this model then proceed by introducing first, a quantity ω_i defined as the fraction of the nearest neighbour sites of a given molecule i which are vacant and secondly, a generalised free volume $j(\omega)$ into the partition function. The important step is finding $j(\omega)$, and various attempts to evaluate this quantity have been made by Cernuschi and Eyring (1939), Ono (1947), Peek and Hill (1950) and Rowlinson and Curtiss (1951). Table 4.2 shows the values of a critical constant obtained by some of these workers, together with *LJD* values and the experimental values for argon.

Table 4.2

	$P_c V_c / R T_c$
Cernuschi and Eyring	0.342
Ono	0.342
Peek and Hill	0.719
LJD Cell Theory	0.591
Experimental Value for argon	0.292

Rowlinson and Curtiss studied the low density region by evaluating the second virial coefficient and found their results to be better than those given by the cell theory.

4.6   THE TUNNEL MODEL

In this model, devised by Barker (1960), the molecules are conceived as moving in straight tunnels whose walls are formed by neighbouring lines of molecules. The arrangement is best described by considering a straight wire on which a number of spherical beads, each representing a molecule of diameter σ, are able to slide freely. Suppose now that a large number of such wires, with their beads, are set up parallel to each other allowing enough room for the beads on each wire to slide freely. If we look *along* the direction of the wires then the most densely packed arrangement that can exist is that in which each wire is symmetrically surrounded by six others. Then the beads on any 'central' wire can move one-dimensionally along a tunnel formed by the beads on the six 'nearest-neighbour' wires. If the central wire is removed, the beads originally on it will still in the main move longitudinally along the tunnel axis, but small motions transverse to this axis will also be possible. Indeed, the diagram in Fig. 4.2 can equally well represent a view down along a tunnel axis and the inner area can then represent a 'free area', A_f, analogous to the 'free volume' of the cell model. A_f is the area available to the centre of each molecule perpendicular to the tunnel axis.

Barker first applied the model to an assembly of N rigid spheres and assumed that the total configurational partition function $Q(N, V)$ is, to a good approximation, the product of the configurational partition functions for longitudinal and transverse motion. He also supposed that the volume V is divided into K tunnels each of length ℓM and containing M molecules. Then, assuming that the tunnels are close-packed, he derived the result

$$Q(N, V) = e^N (\ell - \sigma)^N A_f^N. \tag{4.16}$$

The theory was then developed by considering two approximations for A_f. The first of these is a 'smearing approximation' which gives

$$A_f = \pi(r - \sigma)^2$$

where r is the distance between the axes of neighbouring tunnels. A second, and better, approximation is to write

$$A_f = \pi(r - \sigma')^2$$

where σ' is the average closest distance of approach of a central molecule of diameter σ to the line of centres of similar molecules equally spaced at a spacing ℓ. Using these two approximations the equation of state can be derived.

Barker (1961) later extended the tunnel model to deal with the *LJD* 12-6 potential. The results for the equation of state agree well with experiment at high densities, but not so well at lower densities.

4.7 THE WORK OF KIRKWOOD AND OTHERS: INTEGRAL EQUATIONS

The approach to a theory of the liquid state now almost universally used is due to Kirkwood. He saw that it was mathematically possible to calculate the **radial distribution function** $g(r)$ or number density $\rho(r)$ from first principles. These functions were defined in Chapter 2.3. The local number density of a liquid varies rapidly from place to place because of its discrete molecular structure. Suppose that the centre of a molecule is known to be at a certain place, taken as the origin of coordinates. There is unlikely to be any other molecule whose centre is nearer to the origin than one effective diameter, but in a close-packed arrangement it is geometrically possible to have as many as twelve spheres in contact with a central one. For an irregular assembly which is not quite close-packed we can have a 'shell' containing fewer spheres — about nine for a liquid under ordinary conditions. We can imagine a liquid built up one shell at a time, starting from a single molecule enclosed by larger and larger 'Chinese boxes'. This leads to the idea that, if we travel outwards from the centre of a typical molecule, the local number density is not constant but oscillates about its mean value. The importance of the function $g(r)$ lies in the fact that the thermodynamic properties can be expressed in terms of $g(r)$ and $\phi(r)$.

At the time of Kirkwood's work in 1935 the radial distribution function had already been discovered as a result of X-ray studies of liquid structure. These showed that the spacings between pairs of molecules are distributed in the way suggested above, with certain distances being much more probable than others as a result of the geometrical constraints imposed when attempting to fill up space with spherical molecules. Kirkwood's contribution was to show that this situation can be handled mathematically. The problem of calculating the most likely structure of a liquid is, conceptually, rather like the quantum-mechanical problem of calculating the structure of a large molecule in which a movement of one atom or electron can affect all the others. The mathematical tool used by Kirkwood was an **integral equation**. If we are about to place a molecule in a certain position we must first enquire whether it is too near the central molecule. If not, we must allow for the further possibilities that the proposed location is either already occupied or made untenable by the proximity of two, three or four other molecules. In effect, we have to make such an examination for each point of space and to average over the positions of other molecules in the neighbourhood. This helps us to understand why **integrations** must be involved, and why these integrations involve averaging over the function that we are trying to calculate.

The mathematical details are very complicated and we can only touch upon the various treatments. The problem of calculating a liquid structure given the energy of interaction of two molecules can, in principle, be formulated exactly but approximations always have to be made in practice. Various approximations were tried by various workers. The form of integral equation which works best is due to Percus and Yevick (1958) and the radial distribution function can now be predicted theoretically to an accuracy comparable to that with which it can be measured. Once this distribution function is known in addition to the interaction of two particles, all the equilibrium properties are known also. If the interaction energy between two molecules is known as a function of distance, the total energy of the assembly can be calculated as an appropriate average of the distribution function and the remaining equilibrium properties can be deduced thermodynamically.

All forms of approach based on the radial distribution function agree on the following point. The distribution function for any liquid always closely resembles the distribution function to be expected if the molecules were hard spheres. That is to say its form is mainly settled by the geometrical considerations discussed briefly above while the fact that real molecules attract one another and have slightly 'soft' cores do not affect it very much. This suggests a 'perturbation' approach starting from an imaginary rigid-sphere liquid, whose thermodynamic properties can now be regarded as known as the result of various computer studies. Such perturbation approaches have, in fact, been developed by Barker and others and the results are extremely promising. The rigid-sphere liquid is much closer in structure to a real liquid than are either a modified solid or a dense gas which the older models took as starting points.

4.8 BERNAL'S STATISTICAL GEOMETRY

This situation has been exploited in another way by Bernal and others. As a result of studies of assemblies of ball-bearings and other models they were able to improve on the 'hole' concept. Their picture of an irregularly packed assembly of spheres is based on the geometrical concept of **Voronoi polyhedra**. These are bounded by imaginary planes bisecting at right angles the lines which join each sphere to its neighbours. Bernal's observation was that only a comparatively small number of geometrical types of polyhedron occur under the conditions approaching close-packing that are appropriate for a real liquid. A real liquid is then thought of as an **equilibrium mixture of a few types of polyhedron**, some compact, and some more open. This is obviously much more correct than thinking of a liquid as being like a crystalline solid with a fraction of its lattice sites empty. The relationship between this viewpoint and the distribution function concept is obviously very complicated and details are still being worked out.

REFERENCES

Books

Barker, J. A., (1963), *Lattice Theories of the Liquid State*, Pergamon Press.

de Boer, J. and Uhlenbeck, G. E. (Editors) (1964), *Studies in Statistical Mechanics, Volume II*, North-Holland.

Eyring, H., Henderson, D. and Jost, W., (Editors) (1967), *Physical Chemistry, Volume II*, Academic Press.

Hill, T. L., (1956), *Statistical Mechanics*, Chapter 6, McGraw-Hill Book Company Inc.

Mayer, J. E. and Mayer, M. G., (1940), *Statistical Mechanics,* John Wiley and Sons, Inc., New York.

Moore, W. J., (1964), *Physical Chemistry, 4th Edition*, Longmans.

Pryde, J. A. (1966), *The Liquid State*, Hutchinson University Library.

Rice, S. A. and Gray, P., (1965), *The Statistical Mechanics of Simple Liquids*, Chapter 2, Interscience Publishers.

Rushbrooke, G. S., (1949), *Statistical Mechanics*, Oxford University Press.

Tabor, D., (1969), *Gases, Liquids and Solids*, Penguin Books.

Trevena, D. H., (1975), *The Liquid Phase*, Chapter 5, The Wykeham Science Series.

Papers

Barker, J. A., 1960, *Australian J. Chem.* **13**, 187.

Barker, J. A., 1961, *Proc. Roy. Soc.,* **A259**, 442.

Bernal, J. D. and Mason, J., 1960, *Nature,* **188**, 910.

Bernal, J. D., Mason, J. and Knight, K. R., 1962, *Nature,* **194**, 957.

Bernal, J. D., 1964, *Proc. Roy. Soc.,* **A280**, 299.

Cernuschi, F. and Eyring, H., 1939, *J. Chem. Phys.,* **7**, 547.

de Boer, J., 1952, *Proc. Roy. Soc.,* **A215**, 4.

Kirkwood, J. G., 1935, *J. Chem. Phys.,* **3**, 300.

Kirkwood, J. G. and Boggs, M. B., 1935, *J. Chem. Phys.,* **10**, 394.

Lennard-Jones, J. E. and Devonshire, A. F., 1937, *Proc. Roy. Soc.,* **A163**, 53.

Lennard-Jones, J. E. and Devonshire, A. F., 1938, *Proc. Roy. Soc.,* **A165**, 1.

Levelt, J. M. H. and Cohen, E. G. D., 1964, Chapter B2 in *Studies in Statistical Mechanics,* edited by J. de Boer and G. E. Uhlenbeck, North-Holland Publishing Company.

Longuet-Higgins, H. C. and Widom, B., 1964, *Mol. Phys.* **8**, 549.

Michels, A., Wijker, Hub. and Wijker, H. K., 1949, *Physica,* **15**, 627.

Ono, S., 1947, Mem. Faculty of Eng., *Kyushu Univ.,* **10**, 190.

Peek, H. M. and Hill, T. L., 1950, *J. Chem. Phys.,* **18**, 1252.

Percus, J. K. and Yevick, G. J., 1958, *Phys. Rev.,* **110**, 1.

Rowlinson, J. S. and Curtiss, C. F., 1951, *J. Chem. Phys.,* **19**, 1519.

Walton, A. J. and Woodruff, A. G., 1969, *Contemp. Phys.,* **10**, 59.

Chapter 5
CHANGES OF PHASE

In this Chapter we examine the theromodynamics of phase changes (which are independent of their precise mechanism) and also summarise some detailed models.

5.1 FIRST-ORDER TRANSITION AND CLAPEYRON'S EQUATION

Textbooks on thermodynamics show that g, the Gibbs function per unit mass of substance (the chemical potential), remains a constant during a reversible change occurring at constant temperature and pressure. The phase transitions of melting, vaporisation and sublimation discussed in Chapter 2 are examples of such reversible changes. It follows that if the vapour and liquid phases of the same substance coexist in equilibrium, then their Gibbs function per unit mass (or, if we prefer, per mole) must be equal. Similar remarks hold for the equilibrium between the liquid and solid phases. During these transitions however there are changes in the entropy and volume.

Consider the phase diagram of a simple substance, one which can exist as a solid, liquid and vapour according to the conditions prevailing (see Fig. 2.2). For definiteness let us consider the liquid-vapour equilibrium line RC (the 'transition curve') and discuss how the Gibbs function g for unit mass varies as we move along this line. Note that for unit mass, or for one mole, we use small letters, g, v, s etc. instead of G, V, S, etc. A small change in g is given by

$$\mathrm{d}g = -s\,\mathrm{d}T + v\,\mathrm{d}P \qquad (5.1)$$

(see Appendix 2) so that

$$\left(\frac{\partial g}{\partial T}\right)_P = -s \quad \text{and} \quad \left(\frac{\partial g}{\partial P}\right)_T = v.$$

Let suffices 1 and 2 respectively denoted the phases which are stable on the low and high temperature side of the equilibrium line RC. Then, for a phase change at T and P, we have

$$g_1 = g_2$$

and, for a phase change at $T + dT$ and $P + dP$, if dP and dT are such that we are still on the equilibrium line,

$$g_1 + dg_1 = g_2 + dg_2 .$$

On subtraction, $dg_1 = dg_2$. Using equation (5.1) this gives us

$$-s_1 \, dT + v_1 \, dP = -s_2 \, dT + v_2 \, dP .$$

Hence
$$\frac{dP}{dT} = \frac{(s_2 - s_1)}{(v_2 - v_1)}$$

$$= \frac{\ell}{T(v_2 - v_1)} \tag{5.2}$$

where $\ell = T(s_2 - s_1)$ is the latent heat of vaporisation per unit mass of the transition. Equation (5.2) is known as Clapeyron's equation and the quantities s, ℓ and v can refer to the phase transition of either unit mass or one mole whichever we prefer to work with. This equation gives the variation of vapour pressure with temperature in terms of the other quantities in the equation. Furthermore, since v_2 is always greater than v_1, dP/dT is always positive.

If we had similarly considered the solid-liquid equilibrium line RA we would have obtained the same result, namely

$$\frac{dP}{dT} = \frac{\ell}{T(v_2 - v_1)}$$

where ℓ is now the latent heat of fusion per unit mass. This equation tells us that if $T^\circ K$ is the melting point of a solid, then the change in this melting point caused by pressure change of dP is dT. Furthermore v_1 and v_2 are now the volumes of unit mass of solid and liquid respectively. If the substance expands on melting, as is usually the case, $v_2 > v_1$ and dP/dT is positive. In other words, increasing the pressure raises the melting point. If $v_2 < v_1$ (as with ice or bismuth) melting causes a decrease in volume, dP/dT is negative and increasing the pressure lowers the melting point.

These familiar phase transitions of vaporisation and melting are examples of a so-called **first order transition**. The reason is that the two first order derivatives of the Gibbs function change discontinuously. These two derivatives are first the volume $v = \left(\dfrac{\partial g}{\partial P}\right)_T$, which changes from v_1 to v_2 as the transition

occurs and secondly the entropy $s = -\left(\dfrac{\partial g}{\partial T}\right)_P$, which changes from s_1 to s_2 be-

cause of the latent heat given by $\ell = (s_2 - s_1)T$.

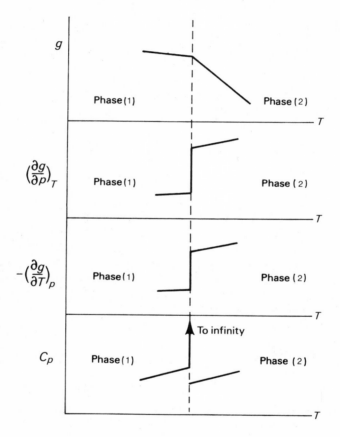

Fig. 5.1   The variation of g, v, s and C_p during a first-order transition.

     The graphs in Fig. 5.1 summarize what occurs during a **first-order transition.** The Gibbs function is continuous but suffers a discontinuous change in gradient. There is a 'jump' in both the specific volume v and the entropy s. The specific heat (C_p) curve shows that the C_p of phase (1) remains finite right up to the transition temperature and then shoots up to infinity during the actual transition itself (when a mixture of two phases is actually present) returning to a finite value for phase (2). The (C_p, T) curve does not show an inclination to 'anticipate' the phase transition by a rapid rise as the transition point in phase (1) is approached. Instead it rises gently as this point is reached and only jumps dramatically to infinity at the actual transition itself, when both phases are present. This is a characteristic of many first-order transitions.

We can explain this behaviour of C_p at the transition as follows. We know that during the transition itself both T and P are constant; thus when P is constant, $dT = 0$ or when T is constant $dP = 0$. Thus

$$C_p = T \left(\frac{\partial S}{\partial T} \right)_P = \infty. \tag{5.3}$$

It also follows that the thermal expansivity, β, and the isothermal compressibility, κ_T, are infinite during the transition since

$$\beta = \frac{1}{V} \left(\frac{\partial V}{\partial T} \right)_P = \infty, \quad \kappa_T = -\frac{1}{V} \left(\frac{\partial V}{\partial P} \right)_T = \infty. \tag{5.4}$$

It must be emphasized that C_p, β and κ_T are infinite only when both phases are present, so that a transition is actually taking place.

To sum up what characterizes a first-order transition: At constant T and P, G remains unchanged. The first-order derivatives of G, $S = -\left(\frac{\partial G}{\partial T} \right)_P$, and $V = \left(\frac{\partial G}{\partial P} \right)_T$, undergo finite jumps. Finally, C_p, β and κ_T become infinite during the transition itself, when the two phases co-exist.

5.2   THE LATENT HEAT OF VAPORIZATION, ℓ_v

The measurement of ℓ_v for liquids with normal boiling points (that is $250°$ to $550°K$) is based on the use of the equation

$$\ell_v = \frac{W}{m}$$

where W is a measured quantity of electrical energy supplied to vaporize a mass m of the liquid. For this purpose a small heating coil is immersed in the test liquid L, which is contained in a small vessel A surrounded by a constant-temperature bath. In this way the test liquid is kept in equilibrium with its vapour. The vessel A is connected to another vessel B outside the bath surrounding A and the temperature of B is kept at a lower value than that of the liquid L. Under these conditions a pressure gradient exists in the vapour of L and so a quantity of L distils over into vessel B. If the rate at which electrical energy is supplied to L is sufficiently small to keep the temperature of A equal to its surroundings, then the whole of the energy supplied is used to vaporize a mass m of the liquid.

For the more interesting cryogenic liquids (see Chapter 14) and for 'simple' liquids with non-polar or weakly polar molecules whose boiling points are $100°$K or less, some data are available in tables by F. Din (ed.) *Thermodynamic Functions of Gases* (Butterworths Scientific Publications, London, 1956). In these tables the corresponding values of ℓ_v, the saturated vapour pressure P, the volumes per mole $v_1{}'$, $v_2{}'$ of the saturated liquid and vapour respectively and the temperature T are tabulated. Zemansky has used these data to plot values of ℓ_v/T_c against $P(v_2{}' - v_1{}')/T$ from about 0.5 T_c to 0.98 T_c, where T_c is the critical temperature. The straight line obtained has a slope of 5.4, that is,

$$\frac{\ell_v/T_c}{P(v_2{}' - v_1{}')/T} = 5.4 \quad (\text{for } 0.5 < \frac{T}{T_c} < 1).$$ (5.5a)

Zemansky shows that this equation can be combined with Clapeyron's equation to give

$$\ln \frac{P}{P_c} = 5.4 \left[1 - \frac{T_c}{T} \right] (\text{for } 0.5 < \frac{T}{T_c} < 1).$$ (5.5b)

which is a relation between reduced temperature and pressure.

A further useful working rule is Trouton's rule, which states that the value of ℓ_v/RT_B, where T_B is the boiling point, is about 9 for many liquids. This can be understood in the following way. The entropy of the vapour does not differ greatly from that of a perfect gas, which is not very sensitive to temperature. The entropy of the liquid is small compared with that of the vapour. Therefore the entropy change on boiling (ℓ/T_B) is also insensitive to temperature. (For further details see Zemansky's *Heat and Thermodynamics*, Chapter 12, 5th edition).

5.3   THE LATENT HEAT OF FUSION, ℓ_F

The most direct way of measuring the latent heat of fusion of a solid is to heat it electrically at a steady rate and plot the (temperature, time) curve, which is of the form shown in Fig. 5.2. The actual phase transition occurs during the constant-temperature region AB of the curve. If m is the mass of solid which melts in time τ then

$$\ell_F = \frac{W}{m} = \frac{EI\tau}{m}$$

where $W = EI\tau$ is the total electrical energy required for the melting process.

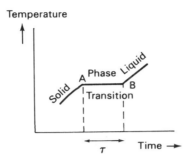

Fig. 5.2 Phase transition of melting.

The melting curves for different substances have been obtained by methods such as the blocked-capillary method (for details see Zemansky, 5th edition). An equation which has been quite successful in representing the shape of such melting curves when the pressures are far above the triple point value is the following, suggested by Simon and Glatzel (1929),

$$\frac{P}{a} = \left(\frac{T}{T_r}\right)^{c'} - 1 .\qquad(5.6)$$

Here T_r is the triple point temperature, a is a constant for a given substance and the exponent c' varies depending on the substance. For example, for the melting curve of solid argon $a = 2240$ atm and $c' = 1.5$. Such a formula can be predicted by various plausible arguments which will not be given here as none of them are rigorous.

5.4 THEORIES OF MELTING

Physicists have been trying to evolve a satisfactory theory of melting for a number of years, but in spite of these attempts no complete theory has so far been developed. The basic problem is this: how do we explain what happens as a solid melts and the atoms go from an ordered phase to a more disordered one? Lindemann (1910) suggested the following explanation. As the temperature of a crystalline solid rises, the individual atoms (or molecules) occupying the lattice sites vibrate with an ever-increasing amplitude determined by the attractive forces between the atom and its neighbours. These are the forces which keep the lattice together. When, however, the amplitude of the vibrations exceeds a certain critical value the effect of these attractive forces is overcome and the lattice structure is destroyed. As Zemansky (1968) puts it: 'In melting, a solid shakes itself to pieces'. Using these ideas Lindemann derived the result

$$\frac{MV^{2/3}\Theta^2}{T_f} = \text{const} \tag{5.7}$$

where M is the molecular weight, V the molar volume, Θ the Debye characteristic temperature and T_f is the melting temperature. This result holds fairly well over wide ranges of T_f for quite a number of substances, both metals and nonmetals. For further details of Lindemann's theory and for subsequent developments, the reader is referred to Ubbelohde's (1965) book *Melting and Crystal Structure* and to Temperley's book on *Changes of State* (1956).

Many attempts were made to show that the models of a liquid discussed in Chapter 4 could predict the melting and vaporisation transition, but the melting transition is still not properly understood. If we use the 'hole' or lattice model of a liquid, likening it to a crystal lattice with vacancies distributed at random, it is fairly easy to show that such an assembly should show a first-order transition at a low enough temperature. The number of vacancies in the lattice should drop discontinuously and at the same time their random distribution throughout the lattice should be modified. The latent heat of fusion is thus associated with the decrease of entropy (increase of crystalline order) associated with this re-arrangement of the lattice. Such theories were undoubtedly on the right lines, but were difficult to compare meaningfully with experiment, as they usually involved both drastic approximations and the existence of several adjustable parameters.

The work of Kirkwood and his school (see Chapter 4.7) on integral equations led to the conclusion that at a certain critical density the solution, the distribution function $g(r)$, should change from the 'damped oscillatory' form characteristic of a liquid to the periodic structure characteristic of a solid. It was predicted that such a change would occur even in an assembly as simple as a gas of rigid spheres without attractive forces. Many physicists were reluctant to accept such a conclusion, but it was finally confirmed in 1957 by computer studies on the rigid sphere assemblies (containing about 100 spheres) carried out at Livermore and Los Alamos. The transition was found to be first order, and to occur at about two-thirds of close-packed density. The jump in density was about 10%, of the same order of magnitude as in real liquids.

One might also ask whether the melting transition is associated with changes in the analytic behaviour of the Mayer virial series (Chapter 4.1.2). It would be tempting to think that the onset of melting occurs at the limit of convergence of the virial series, but unfortunately the situation is not as simple as this, and the matter is under intensive investigation.

Hoover and Ross (1971) have used computer methods to locate the melting transition for assemblies of particles interacting according to various laws of force, and have compared the results with predictions of various melting theories.

5.5   THEORIES OF EVAPORATION

The van der Waals theory has already been described in Chapter 2.2. It should be emphasised that as a theory of evaporation it remains essentially correct, but that the unphysical assumption of an attraction that remains independent of distance can be improved upon at the expense of greatly increased mathematical complication. One possibility is to truncate the virial series (4.5) for a given interaction at the third or fourth term. The resulting approximate equation of state is treated exactly like the van der Waals', while the critical region is located by the condition that $(\partial P/\partial V)$ and $(\partial^2 P/\partial V^2)$ both vanish; one then proceeds as before. Such a relatively simple treatment gives better agreement with experiment than van der Waals.

Possibly the most physically satisfying treatment of a simple liquid available is the model of Longuet-Higgins and Widom already mentioned (Chapter 4.1.3). This does not attempt to explain *analytically* the solidication transition of a rigid sphere gas, but accepts it as a fact. The rigid sphere model is now modified by introducing long-range attractive forces between the molecules. As we have seen, the rigorous consequence of such forces is a correction to the pressure of the type a/V^2, a being constant. They work out the consequences of this model, and find that it leads to a satisfactory over-all representation of the diagram of state of an inert gas. Amongst other things, they are able to locate for this model the triple point at which solid, liquid and vapour are all in equilibrium. Considering the crudity of the representation of both the attractive and repulsive parts of the interaction, the comparison of the predictions of this model with the observed properties of argon are surprisingly good.

Some progress has been made with investigations of the consequences of replacing by something more physically realistic the assumption that the attractive force between two molecules is independent of distance. Application of the Mayer theory (see Chapter 4.1.2) shows that, if the attractive force is a function of distance and if we retain for the moment the assumption that the repulsive forces are of rigid-sphere type, we have to introduce corrections into *all* the virial coefficients and not merely the second, so that the analytic problem becomes far more complicated. However, if the attractive forces do not fall off too rapidly with distance, it is expected (though it has not been formally proved) that the a/V^2 type of correction is still a satisfactory first approximation. Indeed, the good numerical agreeement of the Longuet-Higgins and Widom model with the actual properties of argon seems to bear this out.

Further light has been thrown on this question by some work by Yang and Lee (1952) using the lattice model of an imperfect gas. In this the vessel is supposed divided into cells which may either contain no atoms or not more than one, and all interactions are neglected except those between atoms in neighbouring cells. This is the exact opposite of assuming long-range attractive forces. The consequences of this model can be worked out with complete rigour using results previously obtained by Onsager. A Mayer expansion (Chapter 4.1.2) can be obtained for this model and the onset of liquefaction corresponds precisely with the limit of convergence of the series (4.3) and (4.4). These series are useless for describing the liquid region, but a simple transformation produces series that are valid there. As one might expect the a/V^2 type of correction is found to be a poor approximation for this extreme case of very short range attractions.

If the treatment of the repulsive forces were improved on, one might hope for a model of lattice type that describes both liquefaction and solidification of an imperfect gas. Such models have been studied by Runnels using matrix methods, but they are not at present soluble by closed analytic methods. For details see Chapter 8 of *Phase Transitions and Critical Phenomena,* Vol. 2. This leads us to ask two questions, first how the behaviour of real substance in the critical region may be expected to differ from that of a van der Waals gas and secondly how far it is possible to relate this behaviour to the details of the interactions of real molecules. This has led to the study of so-called *critical exponents*. A critical exponent is obtained by taking some quantity which is becoming very large (or very small) as we approach the critical region. We could then ask how quantities like the specific heat, the compressibility or the difference in densities between the liquid and vapour phases vary with say $T - T_c$ or $V - V_c$.

In the immediate neighbourhood of the critical point this variation is usually found experimentally to be of *power-law* type, hence the name **critical exponent**. The van der Waals theory itself predicts various critical exponents for the van der Waals gas, for example that $|V - V_c|$ should, for both liquid and vapour phases in the critical region, vary as $|T - T_c|^{1/2}$. The *measured* value of the exponent for real substances is much nearer $\frac{1}{3}$ than the above value of ½, and one might look for some connection between this and the effective range of the intermolecular interaction.

In 1965 a conference was held at Washington D.C. at which the critical data for a large number of different types of phase-change (ferromagnetic, paramagnetic, ferroelectric, superconducting, liquid helium etc.) were compared, and it was found that there was often close agreement between comparable critical exponents in these widely different types of system. For example, the spontaneous magnetisation of a ferromagnetic material also behaves roughly like $(T_c - T)^{1/3}$ in the region just below the critical temperature. In fact, it had been already pointed out by Yang and Lee in 1952 that a theory of the critical point of a ferromagnetic could be converted, by appropriate changes of thermodynamic variables, into a simple model of an imperfect gas near its critical region.

It was no surprise that the thermodynamics of apparently distinct types of phase-transition were closely related. The surprising fact was that the corresponding critical exponents seemed to be nearly the same for a 'lattice' type system like a crystalline or metallic magnetic material, and a 'continuum' system like a liquid or solution. This led to the suspicion that one does not need the full details of the intermolecular interaction function to predict the behaviour in the critical region. Indeed, the critical exponents may depend on various weighted averages of this interaction function rather than on its finer details.

We have seen above that we cannot carry through the general Mayer theory of condensation completely because of lack of knowledge of the behaviour of the high-order cluster integrals b_ϱ. However, the above considerations suggest that we may be able to find a limiting form of the theory that should be reliable in the immediate neighbourhood of the critical point. Since 1965 much attention has been paid to search both for relationships between the various critical exponents and for 'scaling-laws', this is to say predictions of these exponents valid in the immediate neighbourhood of the critical point. Past experience of theories of phase-transitions show that the consequences of plausible approximations can be very misleading. Many of the assumptions made in the 'scaling theory' are undoubtedly correct or nearly so, but only a few critical exponents can be calculated on an absolutely rigorous basis at present. The reader is referred to a review article by M. E. Fisher (1967).

5.6 HIGHER ORDER PHASE-TRANSITIONS

We have already considered the characteristics of a first-order transition in Section 5.1, and we obtained a relationship between the latent heat and the change of volume. At the critical temperature these quantities both become zero and it is of interest to ask what types of singularity of the partition function or free energy are then mathematically and physically possible. We choose P and T as independent variables. Since volume $v = (\partial g/\partial P)_T$ and entropy $s = (-\partial g/\partial T)_P$, it is clear that both first derivatives of g, the Gibbs function per unit mass, are finite everywhere. There is, however, no physical reason why one or more of the *second* derivatives of g should not be mathematically singular.

Ehrenfest suggested the possibility that the second derivatives of g might remain finite but be different for the two phases 1 and 2 in equilibrium. Let Δ be an abbreviation for $g_1 - g_2$ and consider a phase change at two adjacent points on the equilibrium line. We have

$$0 = \Delta(P + dP, T + dT) = \Delta(P,T) + dP\,\frac{\partial \Delta}{\partial P} + dT\,\frac{\partial \Delta}{\partial T}$$

$$+ \frac{(dP)^2}{2}\,\frac{\partial^2 \Delta}{\partial P^2} + \frac{(dT)^2}{2}\,\frac{\partial^2 \Delta}{\partial T^2}$$

$$+ dPdT\,\frac{\partial^2 \Delta}{\partial P\partial T} + \ldots$$

Δ is always zero at equilibrium and the small quantities dP and dT have to be determined so that this is so. If $\partial\Delta/\partial P$ and $\partial\Delta/\partial T$ are both finite, equating the sum of the first order terms to zero leads to (5.2). However, they may both be zero, in which case we must look at the second order terms. If $\partial\Delta/\partial P$ and $\partial\Delta/\partial T$ are zero both at (P,T) and at $(P+dP, T+dT)$ we have

$$0 = \frac{\partial \Delta}{\partial P} = \frac{\partial^2 \Delta}{\partial P\partial T}dT + \frac{\partial^2 \Delta}{\partial P^2}dP. \tag{5.8}$$

and
$$0 = \frac{\partial \Delta}{\partial T} = \frac{\partial^2 \Delta}{\partial T^2}dT + \frac{\partial^2 \Delta}{\partial P\partial T}dP. \tag{5.9}$$

Since

$$\frac{\partial^2 g}{\partial T^2} = -\frac{\partial s}{\partial T} = -\frac{C_p}{T}, \quad \frac{\partial^2 g}{\partial P^2} = \frac{\partial v}{\partial P} \quad \text{and} \quad \frac{\partial^2 g}{\partial P\partial T} = \frac{\partial v}{\partial T},$$

equations (5.8) and (5.9) give us two possible relations between discontinuities of specific heat, of compressibility and of coefficient of expansion together with the slope dP/dT of the equilibrium curve.

Comparison of these relations with experiment is, however, difficult. The only certain case of this kind of transition is the transition of a superconducting metal to its normal state.

An alternative possibility is that, for one or both phases, one or more of the second derivatives of g are becoming mathematically 'infinite'. Experimental studies of the critical region of ordinary substances are certainly consistent with a mathematically infinite specific heat and coefficient of expansion. Ultrasonic studies leave it doubtful whether the compressibility actually becomes infinite, but it certainly rises very sharply in the critical region. It is difficult to decide experimentally between a 'discontinuity' and an 'infinity', but the transition in the critical region is now believed to be analytically the same as the so-called lambda transition. This is characterized thus: T, P, G, S and V remain constant and C_P, β and κ_T become infinite. As we shall see in Chapter 14 the (C_p, T) curve has a shape similar to the Greek letter lambda and the transition from liquid helium I to helium II is an exampe of a λ-transition. Another example of a λ-transition is the order-disorder transformation in alloys such as β-brass etc.

We have already mentioned that in a first-order transition C_p does not 'anticipate' the oncoming transition by starting to increase dramatically as the temperature rises to the transition temperature. Instead it remains finite right up to this transition point. In the case of a λ-transition, however, the value of C_p 'anticipates' the transition by starting to rise before the transition temperature is reached (see Fig. 14.3).

5.7 DIFFICULTIES IN STUDYING THE CRITICAL REGION

Since the careful work carried out by Callendar on water it has been known that the finer details of the critical region could be greatly affected by very small quanties of dissolved air. In the 1950's studies by Schneider and other workers showed that the pressure-gradient due to gravity could have a significant effect on a pure substance near the critical region. These facts can be understood in the light of present ideas about the critical region. We have seen that the compressibility becomes very large, if not actually infinite. This means that even the small pressure-gradient associated with gravity can produce an appreciable density gradient and an apparent dependence of the properties on the length of the tube containing the liquid. We also know that a fluid and a magnetic material near their critical regions behave analogously. In this region a magnetic material is extremely sensitive to an applied magnetic field. The effect of a small quantity of magnetic impurity in such a material can be to produce a magnetic domain of dimensions far greater than those of the impurity. In a fluid near its critical point, the introduction of a foreign molecule could produce a modification of density extending over many atomic distances.

These considerations help us to understand both the difficulties of experimenting in the critical region and the commonly observed fact that the scatter of experimental points tends to increase as we approach the critical region. They also justify the rather surprising statement made above that it is difficult to discriminate experimentally between an 'infinity' and a 'discontinuity' in quantities like specific heat.

5.8 LIQUID CRYSTALS

When certain crystalline solids are heated they do not change at once into a liquid but pass first through an intermediate phase known as a **mesophase**. Reinitzer first observed the phenomenon in 1888 by heating solid cholesteryl benzoate, the substance changed first into a turbid liquid at $145°C$ and later into a clear liquid at $179°C$. Since then, several other substances have been found to behave in the same way. The intermediate liquid phase or mesophase is birefringent — a property one associates with a crystal; for this reason the mesophase is usually called a **liquid crystal**. For these substances two transition temperatures are involved — the temperature T_1 at which the crystalline solid melts into the mesophase, and a higher temperature T_2 at which the turbid mesophase changes into a clear liquid.

Most substances which form liquid crystals have long, rod-like molecules. In the solid crystalline phase these molecules are arranged in a way such that there is high spatial and angular correlation. However, in the mesophase the molecules tend to set themselves with their long axes parallel so that the angular correlations, but not the high spatial correlations, of the solid phase still persists. Finally, this angular correlation also disappears in the isotropic liquid phase.

If we consider the mesophase in greater detail it is found that there are various possible structural factors which have led to three basic types of mesophase being identified. A **nematic** mesophase has a low viscosity and is structurally the simplest. In this case the axes of the long molecules tend to arrange themselves parallel to one another, but there is no correlation between the molecular centres. The 'preferred' molecular orientation at a given point in the liquid crystal is described by a vector known as a **director** (de Gennes, 1969). The variations in the orientation of the director inside a nematic mesophase are the cause of the turbid appearance of the phase.

The second type, a **cholesteric** mesophase, is really a special case of a nematic mesophase. If a cholesteric mesophase is sandwiched between two glass surfaces the molecules are arranged in parallel sheets. As we go from one sheet to the next the director is turned through a certain constant angle so that the structure has a helical, repeating pattern. This helical structure has been detected using optical polarization methods.

Thirdly, in a **smectic** mesophase the centres of the molecules are highly correlated in addition to the molecular long axes being parallel. Thus this mesophase has much in common with a solid.

The number and variety of technological applications for liquid crystals are very large and will undoubtedly grow in the future. To mention only a few examples, liquid crystals have been used in (a) electronic display systems, wrist watch faces, portable calculating machines. They have also been used in (b) medical thermography, thermal nondestructive tests of aerospace components and the detection of structural flaws in integrated circuits.

Nematic liquid crystals are used in the type (a) applications just mentioned. These are display systems consisting of a nematic liquid crystal sandwiched between two plates of glass each having a transparent conducting coating. If there is no applied voltage across the plates the enclosed mesophase appears transparent. However the application of a small d.c. or low-frequency a.c. voltage results in the creation of a number of scattering centres in the mesophase. In other words, the mesophase scatters incident light by a process known as an electro-optic effect. If one of the plates is made up of a number of segments each photoetched with a number (0 to 9) or with a letter then the required number or letter may be produced by applying the voltage to the particular segment concerned. In this way numerical displays for electronic clocks, watches and other digital instruments can be produced.

Cholesteric liquid crystals are used in the type (b) applications mentioned previously. In a cholesteric mesophase the 'pitch' of the helical pattern can be caused to change by even very small changes in the temperature. If the pitch changes this will produce changes in the wavelength of any light reflected by the mesophase. Thus this type of mesophase displays abrupt changes in colour due to small temperature changes and gives us a good visual temperature-sensing detector. These methods have been employed to detect structural faults in electronic and aerospace components and in turbine blades. In medicine they have been used to monitor skin temperatures over a fairly large area of the body surface. They thus detect such things as breast tumours, where the temperature would be higher than that of the surrounding area. The method involves applying a light-absorbing coating to the area to be investigated and covering this coating with a calibrated cholesteric liquid crystal. This liquid crystal is designed to respond to a temperature variation of about 4 kelvin over the area examined; the coldest regions appear red or orange and the warmer ones blue or violet. By directly illuminating the area the patient can be examined visually and a colour photograph can be taken as a permanent record.

5.9 GLASSES

There is another interesting class of liquids which form glasses when cooled; such substances are usually polymers of high molecular weight. These liquids can be cooled below their normal melting points to become a supercooled liquid which hardens into a glass after an interval of time. We tend to think of a piece of ordinary glass as being a solid but in fact the molecular arrangement in a glass is very similar to that in a liquid. Indeed X-ray diffraction experiments on glasses produce the pattern of diffuse rings which we associate with a liquid, indicating that there is short-range order but no long-range order. So we can regard the molecular structure in a glass as being an instantaneous photograph of that in a liquid. This glassy state has most of the properties possessed by a liquid: it is isotropic, flows very slowly indeed when stressed and has no well-defined crystal structure or cleavage planes. The only obvious liquid property it does not possess is the ordinary mobility which we always associate with a liquid.

REFERENCES

Books

Domb, C. and Green, M. S. (editors), *Phase Transitions and Critical Phenomena,* Vols. 1 and 2, Academic Press, 1972.

Kallard, T. (editor), *Liquid Crystal Devices,* Optosonic Press, New York, 1973.

Mayer, J. E. and M. G., *Statistical Mechanics,* John Wiley and Sons Inc., New York, 1948.

Pippard, A. B., *The Elements of Classical Thermodynamics,* Cambridge University Press, 1964.

Tabor, D., *Gases, Liquids and Solids,* Chapter 11, The Penguin Library of Physical Sciences, 1969.

Temperley, H. N. V., *Changes of State,* Cleaver-Hume Press Ltd., London, 1956.

Ubbelohde, A. R., *Melting and Crystal Structure,* Oxford University Press, 1965.

Zemansky, M. W., *Heat and Thermodynamics, 5th Edition,* Chapter 12, McGraw-Hill Book Co., 1968.

Papers

Brown, G. H. and Shaw, W. G., 1957, *Chem Rev.,* **57,** 1049.

de Gennes, P. G., 1969, *Molecular Crystals and Liquid Crystals,* **7,** 325.

Fisher, M. E., 1967, *Rev. Mod. Phys.* **30,** 615.

Hoover, W. G. and Ross, M., 1971, *Contemp. Phys.,* **12,** 339.

Luckhurst, G. R., 1972, *Physics Bulletin,* **23,** 279.

Winsor, P. A., 1968, *Chem. Rev.,* **68,** 1.

Yang, C. N. and Lee, T. D., 1952, *Phys. Rev.,* **87,** 404, 410.

Chapter 6
HYDRODYNAMICS AND ACOUSTICS

6.1 THE NATURE OF HYDRODYNAMICS

The literal meaning of the word hydrodynamics is 'motion of water'. However, many of the results that we shall obtain apply to the motion of gases as well as to that of liquids. Hydrodynamics treats a liquid as a **continuous medium,** that is to say it ignores the **atomicity** of real liquids. This leads to no significant error as long as the distances that we are concerned with are large compared with interatomic distances, which are typically of the order of 10^{-10}m. In hydrodynamics, when we refer to a 'particle' of liquid we mean *not* a molecule but a small region of liquid whose density is assumed to be constant, the discontinuous structure of a real liquid being 'smoothed out'. One must remember that hydrodynamics is a very old branch of physics, and was developed by men like Newton, Euler and Daniel Bernoulli. They worked at a time when the concept of 'molecule' was only vaguely appreciated, and nothing was known about its likely size.

Another abstraction that we shall make initially is to neglect 'viscosity'. For example, if two contiguous layers of liquid are sliding over one another we shall assume that there is no force between them. This is never entirely true, but is a good approximation for mobile liquids like water and petrol, although for liquids like oil or glycerine such an assumption cannot be made. A consequence of this assumption is that the pressure, defined in the usual way as force per unit area, in an inviscid liquid is the same in all directions. It should be remembered that in a viscous liquid the pressure is not independent of direction. It has to be represented by a quantity known as a tensor and the mathematics of *viscous* flow is extremely complicated. We derive a few simple results in Chapter 6.8.

6.2 PRESSURE AND 'PARTICLE' VELOCITY

The easiest way of thinking of the 'pressure' in a liquid is that it is one aspect of the forces that molecules of a liquid exert on one another. As an experimental fact we shall assume that the pressure is related to density and temperature through the **equation of state**. The form of the equation of state is determined by the intermolecular forces (see Chapters 3 and 4) but throughout this chapter we shall take it as given. We also need a way of describing the motion of a liquid. We take a small volume, or 'particle', of the liquid, average over the random motions of the actual molecules it contains and assign an average **velocity of translation** to each such 'particle'. If a mass of liquid is moving as a whole (i.e. without relative motion rather like water in a river) this velocity of translation will be practically the same throughout the liquid. However, it is also possible for the velocity to vary from point to point in the liquid, and hydrodynamics describes such non-uniform flow. The velocity of any particle of liquid may vary from instant to instant, and the velocity may also vary from place to place. Velocity is a vector quantity, so to specify completely the motion of a given mass of liquid we need to know the magnitude and direction, or the three cartesian components, of the velocity at each point. We denote it by the vector **v**.

6.3 THE EQUATIONS OF MOTION

We derive the equations of hydrodynamics from two basic laws. First from the conservation of matter, which states that any fluid leaving a given region of space must reappear somewhere else, and secondly from Newton's second law of motion. This states that acceleration is proportional to force. If ρ is the density and (v_x, v_y, v_z) the three components of velocity at any point in space, we can derive the equation of continuity by considering the flow through a small cuboid (fixed in space) of side dx, dy, dz. Figure 6.1 represents this situation.

Fig. 6.1

Because $dy\ dz$ is the area of each of the faces ABCD and EFGH, the rate of flow *out of* the cuboid across the face ABCD is $\rho v_x dy\ dz$ and the corresponding rate of flow *into* the cuboid across the face EFGH is $\rho v_x dy dz$. However these two flows are not in general equal because both ρ and v_x may be different on the planes ABCD and EFGH. These are separated in space by a distance dx. The difference of these two flows, that is the net flow out of the cuboid associated with the x component of velocity, is

$$dy\ dz\ \left[\rho v_{x\,\text{at A}} - \rho v_{x\,\text{at E}}\right]\ .$$

For dx small this is equal to

$$\text{d}x \ \text{d}y \ \text{d}z \ \frac{\partial}{\partial x} \ (\rho v_x)$$

from the definition of a derivative.

Considering now the flow across the planes ADHE and BCGF, we get further contributions to the net flow out of the cuboid associated with the y component of particle velocity. By a similar argument we find dx dy d$z \frac{\partial}{\partial y} (\rho v_y)$ and a third contribution dx dy d$z \frac{\partial}{\partial z} (\rho v_z)$ associated with flow in the z direction. The total of these three terms must be balanced by the rate of change with time of the mass of the liquid contained in the cuboid, which is dx dy d$z \frac{\partial \rho}{\partial t}$.

Thus we have derived the equation of continuity which relates flow velocity and density:

$$\frac{\partial}{\partial x}(\rho v_x) + \frac{\partial}{\partial y}(\rho v_y) + \frac{\partial}{\partial z}(\rho v_z) + \frac{\partial \rho}{\partial t} = 0 \qquad (6.1)$$

which may be written in vector notation

$$\text{div}(\rho\mathbf{v}) + \frac{\partial \rho}{\partial t} = 0 . \qquad (6.1a)$$

We now derive an equation based on Newton's second law. The force on a small volume of liquid is made up of both the pressure (a manifestation of the intermolecular forces) and of external forces. The only external force we shall consider in this book is gravity, though the study of the motion of fluids under the influence of electric and magnetic fields, known as **magnetohydrodynamics,** is a subject of growing importance and interest.

We apply Newton's Law to the motion of a small volume of fluid. In Fig. 6.2 let the two cuboids represent the position of the *same small volume of liquid at two successive instants of time*. By the definitions of v_x, v_y, v_z this small volume has moved through distances v_x dt, v_y dt and v_z dt in the x, y, and z directions respectively. In order to apply Newton's law we want to know how much the three components of velocity *of a given particle* have changed in the time dt. This requires care, because we have chosen to describe the motion of

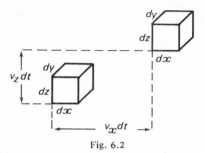

Fig. 6.2

the liquid by specifying the three components of velocity *at each point in space*. As time goes on each element of the fluid moves and the quantity v_x (say) at a given position in space is associated with *different particles of fluid at two successive instants*. It is therefore wrong to say that the acceleration of a given particle is $\dfrac{\partial v_x}{\partial t}$. This is indeed *part* of the acceleration in the x direction, but reference to Fig. 6.2 shows that to calculate the true acceleration of a given particle we must apply a correction. We must allow for the fact that during the time $\mathrm{d}t$ the particle has been transported to a *different position in space* at which the local values of v_x, v_y, v_z may well also be different. Consider first the change in v_x in time $\mathrm{d}t$. It is

$$\mathrm{d}t\,\frac{\partial v_x}{\partial t} \;+\; v_x\,\mathrm{d}t\,\frac{\partial v_x}{\partial x} \;+\; v_y\,\mathrm{d}t\,\frac{\partial v_x}{\partial y} \;+\; v_z\,\mathrm{d}t\,\frac{\partial v_x}{\partial z}$$

where the last three terms give the corrections to the acceleration taking account of the motion of the particle of fluid that we are considering. The force on this small volume is proportional to the gradient of pressure, plus any external force. By definition the forces on the two faces of area $\mathrm{d}y\,\mathrm{d}z$ are both given by $P\,\mathrm{d}y\,\mathrm{d}z$, but P may be different at the two faces because they are spaced at a distance $\mathrm{d}x$. The difference of these two forces is $\delta P = -\dfrac{\partial P}{\partial x}\,\mathrm{d}x\,\mathrm{d}y\,\mathrm{d}z$. The mass of fluid inside the small cuboid is $\rho\,\mathrm{d}x\,\mathrm{d}y\,\mathrm{d}z$.

We have, finally, using force = mass × acceleration that

$$\rho\left[\frac{\partial v_x}{\partial t} + v_x\frac{\partial v_x}{\partial x} + v_y\frac{\partial v_x}{\partial y} + v_z\frac{\partial v_x}{\partial z}\right] = -\frac{\partial P}{\partial x}.$$

Similarly

$$\rho\left[\frac{\partial v_y}{\partial t} + v_x\frac{\partial v_y}{\partial x} + v_y\frac{\partial v_y}{\partial y} + v_z\frac{\partial v_y}{\partial z}\right] = -\frac{\partial P}{\partial y}. \tag{6.2}$$

$$\rho \left[\frac{\partial v_z}{\partial t} + v_x \frac{\partial v_z}{\partial x} + v_y \frac{\partial v_z}{\partial y} + v_z \frac{\partial v_z}{\partial z} \right] = - \frac{\partial P}{\partial z} - \rho g \,,$$

where we are supposing that gravity is the only external force and that it acts downwards in the z direction.

Equations (6.2) can also be written in vector notation:

$$\rho \left[\frac{\partial \mathbf{v}}{\partial t} + (\mathbf{v} \cdot \mathrm{grad}) \mathbf{v} \right] = - \mathrm{grad}\, P - \rho \mathbf{g} \,. \tag{6.2a}$$

Some books use the notion of 'differentiation following a particle' and write $\dfrac{\mathrm{D}\mathbf{v}}{\mathrm{D}t}$ for the terms in square brackets in equation (6.2a). Equations (6.1) and (6.2), together with the fact that P and ρ are related by the equation of state, constitute the equations of hydrodynamics. They are too complicated to solve generally, so we are reduced to studying particular cases. As in other branches of physics there are enough 'standard problems' which can be solved exactly to enable us to find our way about the field, but study of an actual situation nearly always involves considerable numerical work. (Some of the problems which can be solved exactly are also extremely close to reality.)

6.4 THE 'INCOMPRESSIBLE' AND 'ACOUSTIC' APPROXIMATIONS

We shall be mainly concerned with two limiting situations. First the 'incompressible' approximation, which takes the density ρ to be independent of pressure and is a very good description of the motion of liquids in many situations. This is because the compressibility of ordinary liquids is numerically very small. Secondly the 'acoustic' approximation, which is based on the assumption that pressures and densities vary very little from their initial or mean values. This is also physically sensible because the ear is an extremely sensitive instrument which will respond to changes of pressure of a very minute fraction of an atmosphere. In addition, we can safely neglect squares and higher powers of such quantities. It should be stressed that sound *is* the propagation of small pressure differences in a liquid or gas; nothing else is involved. Although the ear does not respond to steady pressures, it is extremely sensitive to varying pressures over quite a wide range of frequencies (from 10 up to about 10^4 Hertz).

We study two main processes in hydrodynamics. These are the bulk flow of liquid, for which it is a very good approximation to treat the density as constant, and the propagation of 'sound', i.e. of differences of density and pressure, from place to place in the liquid. Both of these processes are particular cases of the equations (6.1) and (6.2). It is instructive now to make a count of the number of variables: pressure, density and the three components of velocity.

Thus there are five unknown quantities in all, and between them we have five equations. In addition to the single equation (6.1), there are the three equations (6.2), one for each of v_x, v_y, v_z. Furthermore pressure is related to density through the equation of state. Thus we have five relations between five unknowns.

In the theory of electricity we have the same situation, just enough equations to determine the unknowns. It usually turns out that the initial and boundary conditions appropriate to a physical problem in electromagnetics are enough to fix the solution of the differential equations *uniquely*, and there are many situations in which this uniqueness can be formally proved. In electrostatics, for example, if the electrostatic potential is known at all points of a closed surface in free space we can also determine it at all points inside that surface. The solution of the relevant differential equation is uniquely fixed by the boundary conditions on the closed surface. This is what one would expect from the physics of the problem, and it is in fact confirmed mathematically. One might expect similar considerations to hold in hydrodynamics, but they do not. This seems paradoxical.

A well known example is the flow of liquid through a cylindrical pipe. The physically appropriate 'boundary condition' is that there can be *no flow* perpendicular to the walls of the pipe. However this is not enough to determine the flow, as simple experiments based on the introduction of dye into the fluid show. These demonstrate that we do *not* always get simple *laminar* flow in which the velocity is always parallel to the axis of the pipe. We get laminar flow if the pressure difference producing it is small, but we can get 'wiggly' lines of flow as well. We can also have a situation known as 'turbulence', in which the particles of fluid seem to rush madly about the tube and the motion never settles down at all to a fixed pattern. Clearly none of these modes of flow transgress the 'boundary condition' at the walls of the tube. Thus it is an experimental fact that in hydrodynamics the boundary conditions do *not* determine the flow uniquely.

6.5   IRROTATIONAL MOTION: BERNOULLI'S EQUATION

Lord Kelvin showed that one can impose a further restriction on the flow sufficient to ensure that it *is* determined uniquely by the boundary conditions. He also showed that this so-called **irrotational** flow is the flow which corresponds to the lowest possible total kinetic energy consistent with a given set of boundary conditions. For a proof of this 'minimum energy theorem' the reader is referred to standard books, e.g. Lamb *Hydrodynamics* or Milne-Thomson *Theoretical Hydrodynamics*. It suggests, to any physicist familiar with the proverbial 'meanness' of Nature, that the motion corresponding to the least possible kinetic energy is certainly a reasonable first approximation to what actually happens. And so it turns out. The irrotational solution of a problem always gives preliminary insight into it, and in many cases one need look no further, the actual motion being very close to the irrotational solution. In other instances it must be modified. We shall meet both types of situation in this Chapter.

The irrotational assumption may be stated in various ways. Perhaps the simplest is to assume that there exists a quantity ϕ (a function of x, y, z and time) such that, at every point in the liquid

$$v_x = -\frac{\partial \phi}{\partial x}, \quad v_y = -\frac{\partial \phi}{\partial y}, \quad v_z = -\frac{\partial \phi}{\partial z}. \tag{6.3}$$

A mathematically equivalent statement is

$$\frac{\partial v_x}{\partial y} = \frac{\partial v_y}{\partial x}, \quad \frac{\partial v_x}{\partial z} = \frac{\partial v_z}{\partial x}, \quad \frac{\partial v_y}{\partial z} = \frac{\partial v_z}{\partial y}. \tag{6.4}$$

It is easy to verify that (6.3) implies (6.4) and the reader is referred to books on vector analysis for a formal proof that (6.4) implies (6.3). In vector language, we may write

$$\mathbf{v} = -\text{grad } \phi \tag{6.3a}$$

or equivalently $$\text{curl } \mathbf{v} = 0. \tag{6.4a}$$

At first sight it seems that this restriction on the possible behaviour of the velocity vector $\mathbf{v}$ is extremely drastic, but we shall show that we can nevertheless always find such a solution of equations (6.2). The function ϕ is defined as the **velocity-potential**, and the first step in tackling an irrotational flow problem is to find a suitable form of this function. It gets its name by analogy with electrostatic potential, whose negative gradient is the electrostatic field vector. However the analogy is not perfect, because we can actually measure electrostatic potential in the laboratory but we cannot measure velocity potential. We find that it has very convenient mathematical properties and that its introduction greatly simplifies the calculation of quantities (like particle velocity and pressure) that can be measured. Since it cannot be measured itself one must describe it as an auxiliary quantity. Similar quantities occur in other parts of mathematical physics, for example the vector potential and Hertzian vector in electromagnetics, or the wave-function in quantum mechanics. None of these quantities can be measured by any known experiment and, like the velocity potential, they drop out before the calculation is finished. Nevertheless, their use greatly simplifes the work.

Consider now the first of equations (6.2). Using (6.3) we have

$$\frac{\partial v_x}{\partial t} = -\frac{\partial^2 \phi}{\partial x . \partial t}.$$

Using (6.4) we have

$$\frac{\partial v_x}{\partial y} = \frac{\partial v_y}{\partial x}, \quad \frac{\partial v_x}{\partial z} = \frac{\partial v_z}{\partial x}.$$

Hence this first equation (6.2) becomes

$$-\frac{\partial^2 \phi}{\partial x . \partial t} + v_x \frac{\partial v_x}{\partial x} + v_y \frac{\partial v_y}{\partial x} + v_z \frac{\partial v_z}{\partial x} = -\frac{1}{\rho} \frac{\partial P}{\partial x}.$$

Remembering that P is a known function of ρ because of the equation of state, this whole equation can be integrated with respect to x. We get

$$-\frac{\partial \phi}{\partial t} + \frac{1}{2} (v_x^2 + v_y^2 + v_z^2) = -\int \frac{\mathrm{d}P}{\rho} + G(y, z, t)$$

where G is an arbitrary function. If we treat the second and third of equations (6.2) similarly, we get the same quantity on the left-hand side in all three cases. Comparing the three equations gives some information about the arbitrary functions $G(y, z, t)$ etc. We can conclude that all three of equations (6.2) can be satisfied if we have

$$\int \frac{\mathrm{d}P}{\rho} = \frac{\partial \phi}{\partial t} - \frac{1}{2} v^2 - gz + F(t) \tag{6.5}$$

where $F(t)$ is a function of time only. The term $-gz$ takes account of the effect of gravity, which appears only in the third of equations (6.2). The significance of the arbitrary function $F(t)$ (which may, *but need not*, be zero) is that fluid motion is produced by pressure *differences*. The introduction of a uniform pressure on the whole body of fluid would not affect the motion. Much confusion can be caused if we forget the existence of the $F(t)$ term.

Equation (6.5) is known as **Bernoulli's equation**. Its existence proves the assertion above, which we now amplify slightly. Assume that we can find a velocity potential ϕ that satisfies the boundary conditions and the equation of continuity (6.1). This is always possible. (We shall verify explicitly that this is so in both the 'incompressible' and 'acoustic' approximations later.) We can then calculate the three components of velocity from (6.3) and then the pressure everywhere from (6.5). The fact that equations (6.2) *can* be integrated once is not an accident and is closely related to the existence of an energy integral in dynamics. Indeed, since the equations of hydrodynamics are derived from dynamics, their properties must be similar.

In the particular case where the motion is steady ($\partial\phi/\partial t = 0$ everywhere) equation (6.5) seems to lead to the conclusion that the pressure is least where the stream velocity is greatest. If we have water flowing in a constricted tube, intuition suggests that the pressure should be greatest at the constriction where the velocity is greatest. In fact, experiment soon shows that it is least. This fact is known as **Hawksbee's Law**. In conditions of very rapid flow, e.g. in pumps or near ships' propellers, the pressure can even drop to zero or negative values. Then we can have rupture of the liquid, known as **cavitation** (Chapter 8.3).

Mathematically, the problem of determining possible irrotational flows is decidedly simpler than solving the equations (6.1) and (6.2) directly. This is because the *three* quantities v_x, v_y, v_z are replaced by the *one* quantity ϕ, and we have integrated once to arrive at (6.5). Equations (6.1) and (6.5) simplify even further if we make the incompressible approximation, that is, we assume that the density is constant and independent of pressure. This is a satisfactory approximation in many situations because of the very low compressibilities of liquids. If we take ρ as constant and substitute (6.3) in (6.1) we find

$$\frac{\partial^2\phi}{\partial x^2} + \frac{\partial^2\phi}{\partial y^2} + \frac{\partial^2\phi}{\partial z^2} = 0 \qquad (6.6)$$

while (6.5) becomes

$$\frac{P}{\rho} = \frac{\partial\phi}{\partial t} - \tfrac{1}{2}v^2 - gz + F(t). \qquad (6.7)$$

Equation (6.6) is identical with Laplace's equation for the electrostatic potential (or for gravitational potential, or *any* potential of 'inverse-square law of force' type). Therefore any problem in electrostatics has an analogue in hydrodynamics, in the sense that we can use solutions of Laplace's equation in either context. However, Bernoulli's equation (6.7) has no analogue in electrostatics.

6.6 SOME EXAMPLES OF IRROTATIONAL AND INCOMPRESSIBLE FLOW

We shall now show that very simple solutions of Laplace's equation can lead to interesting situations in incompressible hydrodynamics.

6.6.1 The oscillating bubble

Consider the solution $1/r$ corresponding to an isolated electric charge, and assume a solution $\phi = A(t)/r$. Plainly, the gradient of ϕ will be in the r

direction since ϕ does not vary with the polar angles. This solution therefore corresponds to *radial* motion and can describe the motion of water in the neighbourhood of an expanding or contracting spherical bubble. Let the radius of the bubble at time t be $a(t)$. The velocity of the liquid just outside the bubble must be da/dt. The radial velocity corresponding to the above form of ϕ is, by

(6.3a), $-\dfrac{\partial \phi}{\partial r} = \dfrac{A(t)}{r^2}$. If we put $A(t) = a^2 \dfrac{da}{dt}$, the velocity in the fluid takes

the required value da/dt at the surface of the bubble (given by $r = a$). On substituting this value of ϕ in equation (6.7) we obtain the following expression for the pressure in the water in the neighbourhood of the bubble

$$\frac{P}{\rho} = \frac{1}{r}\frac{d}{dt}\left[a^2\frac{da}{dt}\right] - \frac{a^4}{2r^4}\left[\frac{da}{dt}\right]^2 + F(t) . \tag{6.8}$$

For r large the first two terms are negligible, and we conclude that $F(t) = P_\infty/\rho$, where P_∞ is the pressure in undisturbed water. Let the pressure inside the bubble be $P(a)$. This value will be small if the bubble is a vapour cavity, but can be very large if the bubble is produced by an explosion. We have a further boundary condition that, at $r = a$, the pressure in the water and the pressure inside the gas bubble must be equal if surface tension is negligible. Putting $r = a$ in (6.8) and simplifying, we find

$$\frac{P(a) - P_\infty}{\rho} = a\frac{d^2a}{dt^2} + \frac{3}{2}\left[\frac{da}{dt}\right]^2 , \tag{6.9}$$

which enables us to calculate the radius a of the bubble as a function of time if $P(a)$ is known. We have thus arrived at an equation of motion for the radius a. This equation has to be solved numerically but the solution is always of the

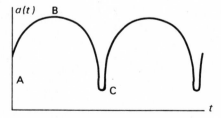

Fig. 6.3 Variation of bubble radius with time.

general type shown in Fig. 6.3. If we start at a point like A, the solution is a fair representation of the later stages of an underwater explosion, for which we can safely neglect the compressibility of the water. If we start at a point like B, we have a description of a collapsing cavity, in which the internal gas pressure is small. Once we have determined $a(t)$ we can substitute into (6.8) and determine the pressure in the liquid in the immediate neighbourhood of the bubble. Because of the large values of d^2a/dt^2 we conclude that the pressure can attain extremely high values in the neighbourhood of a bubble minimum such as C in Fig. 6.3. This throws light on the very destructive effect of cavitation on propellor blades and in pumps by showing that the very high pressures in the neighbourhood of collapsing vapour cavities can exceed the yield stress of most metals.

It is interesting to see that we can integrate equation (6.9) once to get an equation of energy balance. Multiply it through by $4\pi a^2 \dfrac{da}{dt}$. Then since $V = 4\pi a^3/3$ is the volume of the bubble, it will be found that (6.9) can be integrated with respect to t to give

$$\int P(a)\, dV - P_\infty V = 2\pi a^3 \left[\frac{da}{dt}\right]^2 \rho + \text{constant}. \qquad (6.10)$$

The two terms on the left-hand side correspond respectively to the work done by the gas in the bubble in expanding and the work done in pushing the water outwards against the ambient pressure P_∞. We conclude that the term $2\pi a^3 \left[\dfrac{da}{dt}\right]^2 \rho$ represents the kinetic energy of the water. This can in fact be verified by integrating the expression $\frac{1}{2}\rho \left[\dfrac{\partial \phi}{\partial r}\right]^2 dV$ over the region outside the bubble for all values of r between a and infinity.

This leads to the concept of **hydrodynamic virtual mass**. Any object moving in a fluid moves part of the fluid and the extra kinetic energy can, if desired, be attributed to an apparent extra mass of the body. In this particular case the 'virtual mass' is $4\pi\rho a^3$, three times the mass of liquid displaced by the bubble. The form of the curve in Fig. 6.3 can now be qualitatively understood; the bubble is behaving like a body of large mass when a is large and of very small mass when a is small. Hence the very rapid 'turn-round' near the minima. Hence we also get large values of d^2a/dt^2 and corresponding large pressures near a collapsing cavity of vapour, according to equation (6.8). A very similar phenomenon can occur with underwater explosion bubbles. Several oscillations of the type shown in Fig. 6.3 can occur at large depths and at each minimum of the bubble another peak is generated in the pressure in the water outside.

Thus this very simple solution of Laplace's equation has produced some very interesting consequences. This irrotational solution has given a good insight into the problem and need not be modified. The analogue of the electrostatic problem of a sphere in a uniform electric field produces conclusions of equal interest about the disturbance of uniform flow by the presence of an obstacle. The irrotational solution of this problem also gives a good initial insight, but as we shall see below it needs some modification.

6.6.2 Flow in the neighbourhood of a sphere

For a uniform flow along the z direction with velocity U the velocity potential is simply $-Uz$. As in electrostatics, the effect of a sphere can be represented by adding to this the potential of a 'dipole' or 'doublet' located at the centre of the sphere. This potential is z/r^3, which can be verified by actual differentiation to be a solution of Laplace's equation. Therefore the corresponding incompressible flow automatically satisfies the continuity equation (6.1).

It is more usual to work in spherical polar coordinates, so we take our origin at the centre of the sphere and our line $\theta = 0$ along the z axis. We now consider a movement of fluid described by the velocity potential

$$\phi = -Ur\cos\theta + \frac{B\cos\theta}{r^2}.$$

At a great distance from the sphere, this reduces to a uniform velocity in the z direction. If the radius of the sphere is a, our boundary condition is that, at $r = a$, the radial velocity is zero. We obtain the radial velocity by differentiating ϕ with respect to r (see equation (6.3)) and we can choose B so that $\partial\phi/\partial r = 0$ at $r = a$ for all values of θ. That is to say that the velocity potential appropriate to our problem is

$$\phi = -Ur\cos\theta - \frac{Ua^3\cos\theta}{2r^2}. \tag{6.11}$$

This certainly describes a flow satisfying the equation of continuity and the boundary conditions at the surface of the sphere and at infinity. At each point in the fluid we can, by differentiating, calculate the three components of velocity. Alternatively we can use equation (6.3) in the form $v_r = -\partial\phi/\partial r$, $v_\theta = -\dfrac{1}{r}\dfrac{\partial\phi}{\partial\theta}$. The directions of flow at each point are shown in Fig. 6.4.

Fig. 6.4 Irrotational flow in the neighbourhood of a sphere.

The effect of the sphere may be described as an 'opening out' of the flow-lines of the undisturbed uniform flow. (We notice the symmetry). We can calculate v_θ at each point near the surface, hence the pressure at each point from Bernoulli's equation (6.7). If we do this, we conclude that the total net force on the sphere is zero. This is contrary to experiment and is known as **d'Alembert's paradox**. A similar result always occurs if the flow *everywhere* round an obstacle is taken to be irrotational. In reality, solution (6.11) implies considerable *shearing* of the fluid, particularly at regions like A and B in Fig. 6.5.

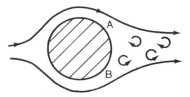

Fig. 6.5 Actual flow in the neighbourhood of a sphere.

The result is that **vortices** are created at such regions and are progressively shed downstream, the eddying region behind the obstacle being known as the **wake**. Study of conditions in regions such as A and B is a separate branch of hydrodynamics known as **boundary layer theory**, although outside the wake solution (6.11) is still a good approximation. For such a situation d'Alembert's paradox does not apply, and there can be a finite force on the obstacle. We can determine this law of force as follows. The water in the wake is nearly at rest relative to the obstacle so the pressure in this region is practically constant. At the front of the obstacle fluid is continually being brought up to the neighbourhood of the obstacle and, in this neighbourhood, its direction of flow is changed (Fig. 6.5).

This is a problem similar to that of calculating the force exerted on a wall by a steady jet. The momentum of the fluid is destroyed by the wall and the water then runs up or down it. For a jet of unit cross-sectional area the momentum content is ρU per unit length, while the volume that is brought up to the wall in unit time is U per unit area. Therefore such a jet exerts a force ρU^2 on the wall. In the geometry of Fig. 6.5 the situation is similar, but not all the momentum in the direction of U is destroyed. However, the rate at which momentum is brought up and destroyed is still proportional to ρU^2. We thus expect the force on an obstacle to be proportional to ρU^2. In fact it is commonly expressed as $\frac{1}{2}\rho U^2 C_D A$, where A is the area of cross-section of the object and C_D (known as the **drag coefficient**) depends on the shape of the obstacle. For a sphere it is of the order of $1/3$, while for a very flat-nosed obstacle it can even be greater than 1. There seems to be no particular reason for including the factor $\frac{1}{2}$ in the above formula, but by convention the drag coefficient is always defined in this way. The above argument is due to Newton, and the above formula is often described as **Newton's law of force**.

In fact the pattern of flow, and consequently the drag coefficient C_D, can change slightly as U increases, but the above formula holds over a very wide range of values of U. We meet the concept of hydrodynamic virtual mass here too. If the fluid motion produced by a moving sphere were completely irrotational, the additional kinetic energy due to motion of the fluid would correspond to *half* the mass of fluid displaced by the sphere.

6.6.3 The rising gas-bubble

Equation (6.11) can also describe the flow round a bubble of gas rising steadily through a liquid. Experimentally it is found that its profile resembles a 'mushroom cap' and that it is nearly spherical at the top. The velocity potential (6.11) describes a flow that is tangential at all points on the surface of a sphere of radius a. If we reverse the sign of U, we get a downward flow past the bubble. Then we can take coordinate axes fixed relative to its centre of curvature, which we take as the origin in polar coordinates.

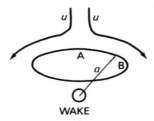

Fig. 6.6 Flow round a rising gas-bubble.

Let a be the radius of curvature. The solution (6.11) ensures that the flow is tangential to the sphere. It was pointed out by Sir Geoffrey Taylor that, if we assume the right relation between U and a, the pressure in the liquid can be made the same at points A and B just outside the bubble. It clearly is the same everywhere *inside* the bubble, being equal to the gas pressure. Just outside the bubble the radial velocity v_r is zero, and the tangential velocity v_θ is given, according to (6.3), by $-\dfrac{1}{r}\dfrac{\partial\phi}{\partial\theta} = \dfrac{3U}{2}\sin\theta$ (after differentiating (6.11), putting $r = a$ and changing the sign of U.) Inserting this into Bernoulli's equation (6.7) and putting $\partial\phi/\partial t = 0$ and $F(t)$ a constant for steady flow, we find that the pressure is given as a function of θ by

$$\frac{P}{\rho} = -\frac{9}{8}U^2\sin^2\theta - ga\cos\theta + F(t).$$

It is observed in practice that the semi-vertical angle of the cap is about $20°$, that is, small compared with a radian. At A in Fig. 6.6 $\theta = 0$, and we can choose the constant F to match the pressure in the liquid at A. For θ small, $\sin\theta$ is nearly equal to θ and $\cos\theta$ is nearly equal to $1 - \dfrac{\theta^2}{2}$. Therefore, the condition that the pressure in the liquid should be constant over the portion of the bubble surface in the immediate neighbourhood of A is

$$U^2 \;=\; \frac{4}{9}ga. \tag{6.12}$$

This relation has been verified in practice for a wide variety of sizes of bubble ascending through water. It also applies to the ascent through the atmosphere of the 'mushroom cloud', the hot gases produced by an atomic bomb. Again we find that quite a simple solution of Laplace's equation leads to physical results of real interest in two quite distinct situations: flow around a solid obstacle and flow around a rising gas-bubble.

6.6.4 Surface waves on a liquid

Another situation in which incompressible and irrotational hydrodynamics works well is in the discussion of surface waves on a channel of uniform depth. We suppose that, at $t = 0$, the free surface has the form $z = z_0 \cos kx$ and we ask under what condition such a disturbance can be propagated in the x direction without change of form. It turns out that this is possible provided that the amplitude of the disturbance is small compared with its wave-length and the depth of the channel.

We let $z = 0$ represent the undisturbed surface and let the deformed surface be given as a function of x and t by

$$z \;=\; z_0 \cos[k(x - Vt)] \ . \tag{6.13}$$

Since we are discussing *plane* waves, there will be no variation with y. This represents a surface deformed into a sinusoidal wave which is being propagated with a velocity V. We shall show that we can satisfy the hydrodynamic equations, the boundary conditions and the irrotational assumption, provided that we choose the correct relationship between V and k, and provided also that we neglect quantities of the order z_0^2; k is $2\pi/\lambda$ where λ is the wave-length, and so kV is 2π times the frequency. V will vary with k (or the wavelength) and the relationship is known as a dispersion relationship, by analogy with optics. Differentiating (6.13) with respect to t, we have

$$\frac{\partial z}{\partial t} \;=\; kVz_0 \sin[k(x - Vt)] \ . \tag{6.14}$$

We now look for a velocity potential that corresponds to such a normal component of velocity at $z = 0$. It must vary like $\sin[k(x-Vt)]$ with x and t, and it must be a solution of Laplace's equation if we are to satisfy the equation of continuity. If the channel is of great depth, the appropriate velocity potential is

$$\phi = -Vz_0 e^{kz} \sin[k(x-Vt)] . \tag{6.15}$$

Differentiating with respect to x and z, we easily verify that it satisfies Laplace's equation (6.6). For the z component of velocity, we find

$$-\frac{\partial\phi}{\partial z} = kVz_0 e^{kz} \sin[k(x-Vt)] . \tag{6.16}$$

If we insert z as given by (6.13) in this, we find that (6.14) and (6.16) only differ by a quantity of the order z_0^2, which we are agreeing to neglect. To this approximation, (6.15) *is* a velocity potential consistent with the assumption (6.13). We could also satisfy Laplace's equation with a solution involving e^{-kz} but, since this would increase with increasing depth, it would be physically inappropriate for a very deep channel.

For a channel of finite depth h, we can replace (6.15) by

$$\phi = (Ae^{kz} + Be^{-kz}) \sin[k(x-Vt)] \tag{6.17}$$

and (6.16) by

$$-\frac{\partial\phi}{\partial z} = (-Ae^{kz} + Be^{-kz})k \sin[k(x-Vt)] . \tag{6.18}$$

At $z = -h$, the vertical component of velocity must vanish at the bottom of the channel for all values of x and t, that is to say we must have

$$B = Ae^{-2kh} . \tag{6.19}$$

At the free surface we can make (6.18) agree with (6.14), and it is again sufficiently accurate if we put $z = 0$ in both exponentials. We conclude that

$$A + B = -Vz_0 . \tag{6.20}$$

A and B are determined from (6.19) and (6.20) and we can determine the appropriate velocity potential for the problem by substituting them into (6.17). It is readily verified that (6.17) reduces to (6.15) for large h, and that, like (6.16), (6.18) agrees with (6.14) if we neglect terms of order z_0^2.

We determine the dispersion relationship between V and k from the condition that the pressure everywhere along the top of the wave-form is atmospheric. If there is no surface tension, the pressure immediately below the surface is also constant. However, if the surface tension γ is finite the pressure in atmosphere and liquid differ by γ/R.

R is the radius of curvature of the surface and is given by the usual formula

$$\frac{1}{R} = \frac{\partial^2 z}{\partial x^2} \Bigg/ \left[1 + \left(\frac{\partial z}{\partial x}\right)^2\right]^{3/2}$$

which, for a surface of the form (6.13), only differs from $\partial^2 z/\partial x^2$ by terms of the order of z_0^2.

We calculate the pressure in the liquid as before from the Bernoulli equation (6.7), after substitution into it of the velocity potential (6.17). The pressure just below the surface must be equal to the atmospheric pressure corrected by γ/R or $\gamma\,\partial^2 z/\partial x^2$. The term $\frac{1}{2}v^2$ in (6.7) always involves z_0^2 and can be neglected because v is of order z_0. However we must *not* neglect the term $\partial\phi/\partial t$, because even though the motion is steady we have chosen to use coordinates such that the wave-form (6.13) is moving relative to them, and therefore $\partial\phi/\partial t$ is *not* zero. Inserting (6.17) into (6.7), we have at the surface

$$\frac{P}{\rho} = -(A + B)kV\cos[k(x - Vt)] - gz + F(t)$$

$$= [-(A + B)kV - gz_0]\cos[k(x - Vt)] + F(t)$$

(6.21)

(using (6.13) for z). This must, at all values of x and t, be equal to the pressure of the atmosphere plus the surface tension correction which, because it is $\gamma\dfrac{\partial^2 z}{\partial x^2}$, is found from (6.13) to be $-\gamma k^2 z_0\cos[k(x - Vt)]$. Therefore we can satisfy the boundary conditions for the pressure by choosing $F(t)$ to be equal to $1/\rho$ times the atmospheric pressure and equating coefficients of $\cos[k(x - Vt)]$. Solving (6.19) and (6.20) for A and B, we find that the boundary condition for pressure can be satisfied at the surface for all x and t if

$$V^2 = \left[\frac{g}{k} + \frac{\gamma k}{\rho}\right]\tanh(kh),$$

(6.22)

which determines the velocity of propagation of waves of small amplitude. Notice that V is quite a complicated function of k, or the wavelength, so that we indeed have **dispersion**. We have derived equation (6.22) using the rather artificial concept of velocity potential. Other derivations are possible, but are usually much longer.

We now look at some special cases of (6.22). The hyperbolic tangent function is well behaved, approaching unity for kh large, while for kh small it behaves like $kh - \dfrac{(kh)^3}{3} + \dots$.

(i) kh small: '**Long waves**'. (Surface tension is always negligible). This situation can occur in a shallow canal, but by far the most important example is the propagation of the **tides**. (6.22) reduces simply to $V^2 = gh$, that is to say **dispersion is absent but the velocity of propagation depends on depth**. The 'tides' are simply the changes of level of the sea produced by the combined effect of the rotation of the earth and the gravitational effects of the sun and moon. Thus their 'period' is of the order of 24 hours and their 'wavelength' of the order of thousands of miles. This is far greater than the depth of any sea, so kh is always very small. The prediction of their times and heights is based on the above considerations, but is a highly technical matter.

(ii) kh large: '**Deep water waves**'. For k small, surface tension is negligible but dispersion is always present. We notice that the velocity of propagation is largest for very long wavelengths. At sea, the disturbances produced by a distant storm travel at different speeds according to the above equation, which reduces here to $V^2 = g/k$. That is to say, the first indication of the approach of a storm can be a disturbance of very long wavelength (small k), often described as a 'swell'.

For wavelengths of the order of a few centimetres, surface tension becomes important. It will be noticed that V^2 attains a minimum value when $g = \gamma k^2 / \rho$ and the corresponding value of k for water corresponds to a wavelength of about 1.7 cm. The distinction is sometimes drawn between 'ripple', with a smaller wavelength than the critical value, and 'wave' of larger wavelength. However the two merge into one another and the distinction is a matter of convenience only. **Rayleigh's** method of measuring surface tension is based on measuring the velocity of propagation of 'ripples'.

In conclusion, we notice the large number of places in which we neglect terms of order z_0^2. This does not diminish the insight provided by the results because z_0 very often *is* negligible. Much work has been done on finite amplitude waves (see Lamb *Hydrodynamics* or Milne-Thomson *Theoretical Hydrodynamics*) but the situation becomes extremely complicated. Except in special cases finite amplitude waves cannot propagate without change of form. In certain circumstances we can have practically discontinuous effects, like 'tidal bores' in the Severn and Trent or the 'hydraulic jump' in fast-flowing rivers, but the reader is referred to advanced works for their study.

The irrotational assumption can nearly always be made for waves in water. For shallow water waves vortices *can* be generated near the bottom, but they are usually localised and of little practical importance.

6.7   THE ACOUSTIC APPROXIMATION: SOUND WAVES

We now leave incompressible hydrodynamics and examine how we can describe propagation of density or pressure differences, that is to say, of 'sound'. We shall be concerned with pressures and densities that differ only slightly from their mean values, these differences being proportional to the amplitude of the sound. The components of particle velocity v_x, v_y, v_z will also be proportional to the amplitude. We neglect all powers and products of such quantities (e.g. the term $v_x \dfrac{\partial \rho}{\partial x}$ which occurs in equation (6.1) is negligible) but the term $\rho \dfrac{\partial v_x}{\partial x}$ is only of the *first* order in the amplitude and must be retained. We shall also adopt a very simple form for the equation of state, using the relation between P and ρ in the form

$$P - P_0 = c^2(\rho - \rho_0).$$
(6.23)

Here c^2 depends on the compressibility of the substance. We shall ultimately identify it with the square of the velocity of sound. The equation of state of a real substance also involves higher powers of $\rho - \rho_0$, but we shall neglect these for small amplitudes of sound and treat $c^2 = \partial P/\partial \rho$ as a constant. The derivative is nearly always **adiabatic** i.e. entropy is constant. See *Gases, Liquids and Solids, pp. 62-64*, D. Tabor (Penguin).

Neglecting second-order terms in the equation of continuity (6.1) we find

$$\rho \left[\frac{\partial v_x}{\partial x} + \frac{\partial v_y}{\partial y} + \frac{\partial v_z}{\partial z} \right] = - \frac{\partial \rho}{\partial t}$$
(6.24)

or
$$\operatorname{div} \mathbf{v} = - \frac{\partial}{\partial t} (\ell n\, \rho)$$
(6.24a)

Similarly we neglect second order terms in the acceleration equation and we find

$$\frac{\partial v_x}{\partial t} = - \frac{1}{\rho} \frac{\partial P}{\partial x}$$

$$= - \frac{c^2}{\rho} \frac{\partial \rho}{\partial x}$$
(6.25)

using (6.23), with similar equations for v_y and v_z, which we may combine in vector form thus

$$\frac{\partial \mathbf{v}}{\partial t} = -c^2 \operatorname{grad}(\ell n \, \rho) \,. \tag{6.25a}$$

Note that the effect of gravity is always small compared with that of the pressure-gradients that occur in a sound-wave. We can eliminate either $\mathbf{v}$ or $\ell n \, \rho$ between equations (6.24a) and (6.25a). To eliminate $\mathbf{v}$, differentiate (6.24a) with respect to t and take the divergence of (6.25a). This gives

$$\nabla^2 (\ell n \, \rho) = \frac{1}{c^2} \frac{\partial^2}{\partial t^2} (\ell n \, \rho) \,. \tag{6.26}$$

To eliminate $\ell n \, \rho$, take the gradient of (6.24a) and differentiate (6.25a) with respect to t.

$$\operatorname{grad} (\operatorname{div} \mathbf{v}) = \frac{1}{c^2} \frac{\partial^2 \mathbf{v}}{\partial t^2} \,. \tag{6.27}$$

Identically, we can show that, for any vector $\mathbf{v}$,

$$\operatorname{grad} (\operatorname{div} \mathbf{v}) = \nabla^2 \mathbf{v} + \operatorname{curl} (\operatorname{curl} \mathbf{v}) \,.$$

If we make the assumption of irrotational motion that curl $\mathbf{v} = 0$, (6.27) also reduces to the wave-equation

$$\nabla^2 \mathbf{v} = \frac{1}{c^2} \frac{\partial^2 \mathbf{v}}{\partial t^2} \,. \tag{6.28}$$

If the motion is irrotational, a velocity potential ϕ must exist. We can show that this also satisfies the wave-equation. We write the original Bernoulli equation (6.5) in the form

$$\int \frac{dP}{\rho} = \frac{\partial \phi}{\partial t} + F(t) \tag{6.29}$$

where we have neglected the small term $\tfrac{1}{2}\mathbf{v}^2$, and the effect of gravity. The arbitrary function $F(t)$ simply corresponds to imposing a spatially uniform pressure on everything, so we can neglect it in this context. Differentiating (6.29) with respect to t

$$\frac{\partial^2 \phi}{\partial t^2} = \frac{1}{\rho} \frac{\partial P}{\partial t} = c^2 \frac{\partial (\ell n \, \rho)}{\partial t} \,. \tag{6.30}$$

We may write (6.24) in terms of ϕ (after dividing through by ρ and putting $v_x = -\dfrac{\partial \phi}{\partial x}$ etc.)

$$\nabla^2 \phi = \frac{\partial}{\partial t}(\ell n\, \rho) = \frac{1}{c^2} \frac{\partial^2 \phi}{\partial t^2} . \qquad (6.31)$$

Thus we conclude that *all* the quantities $\ell n\, \rho$, the three components of the particle velocity $\mathbf{v}$ and the velocity potential ϕ obey the wave equation. The quantity $\ell n\, \dfrac{\rho}{\rho_0}$ is sometimes called the condensation. If ρ never differs greatly from its mean value ρ_0, we have

$$\ell n\, \rho = \ell n\left[1 + \frac{\rho - \rho_0}{\rho_0} \right] + \ell n\, \rho_0$$

$$= \ell n\, \rho_0 + \frac{\rho - \rho_0}{\rho_0} + \text{higher order terms.}$$

Since $\ell n\, \rho_0$ is a constant and $\ell n\, \rho$ satisfies (6.26), the quantity $\ell n\left[\dfrac{\rho}{\rho_0}\right]$ will also satisfy (6.26). Therefore, so does $(\rho - \rho_0)/\rho_0$ the condensation or relative change of density, often denoted by s. Finally, $P - P_0$ is proportional to $\rho - \rho_0$ according to (6.23), so $P - P_0$ (the departure of P from its mean value P_0) also satisfies the wave-equation. We thus have a large number of different quantities which satisfy the wave equation, and it is usually sufficient to determine one of them. We shall work with the velocity potential ϕ.

The wave-equation appears in every branch of mathematical physics in some form, so a great deal of work has been done on methods of finding solutions of it appropriate to many types of initial and boundary conditions. For an extensive discussion of solutions of the wave-equation appropriate to acoustics the reader is referred to Rayleigh's *Theory of Sound*. Here we shall consider two types of solution only, those appropriate to plane waves and those appropriate to waves diverging symmetrically from a point-source.

For a plane wave we have no variation of anything in the y and z directions, so (6.31) reduces to

$$\frac{\partial^2 \phi}{\partial x^2} = \frac{1}{c^2} \frac{\partial^2 \phi}{\partial t^2}, \qquad (6.32)$$

the general solution of which is known to be

$$\phi = F(x - ct) + f(x + ct) \qquad (6.33)$$

where F and f are arbitrary functions. For a spherically symmetrical wave, ϕ varies with r only and the relevant terms of the ∇^2 operator give us

$$\frac{\partial^2 \phi}{\partial r^2} + \frac{2}{r} \frac{\partial \phi}{\partial r} = \frac{1}{c^2} \frac{\partial^2 \phi}{\partial t^2}. \qquad (6.34)$$

This can be reduced to the form (6.32) if we make the substitution $\phi = w/r$. Performing the differentiations we find that (6.34) becomes $\dfrac{\partial^2 w}{\partial r^2} = \dfrac{1}{c^2} \dfrac{\partial^2 w}{\partial t^2}$,

so we conclude that the general solution of (6.34) is

$$\phi = \frac{1}{r} \left[G(r - ct) + g(r + ct) \right] \qquad (6.35)$$

where G and g are arbitrary functions. In fact we are usually concerned with sinusoidal oscillations, i.e. the arbitrary functions will usually be sines and cosines.

Thus the mathematics is especially simple for these two important cases. Note however that there is *no* analogue of (6.35) for the *cylindrically* symmetric case. For discussions of more complicated solutions the reader is referred to Rayleigh's *Theory of Sound*.

Like other branches of mathematical physics, acoustics consists of finding solutions of the equations that satisfy the boundary conditions imposed by the physics of the problem. We have already met these boundary conditions:

(a) In the absence of surface tension, or if the boundary between two media is flat, the pressure is continuous at the boundary.

(b) The normal component of particle velocity is continuous through a boundary.

(a) is a consequence of the definition of pressure while (b) is an obvious consequence of geometry — if one medium is moving normally to the boundary surface, the other medium is being pulled or pushed and necessarily moves in that direction with the same speed. In particular, if a boundary surface is rigid there can be no component of normal flow.

6.7.1 The creation of sound in a fluid medium

We now consider the problem of the *creation* of sound in a liquid or gas by the motion of a solid boundary. This is a most important mechanism, applying to the oscillation of a wire or bell or tuning fork for example, or to the electrically driven oscillations of a loudspeaker element or of a magnetostriction or piezo-electric device. Indeed it applies to any situation in which the motion of a solid is imparting some of its energy to the medium. We shall show that use of the wave-equation, together with the boundary conditions of continuity of pressure and normal velocity, enables us to follow this process in detail for a simple one-dimensional geometry. The reader is referred to Rayleigh's *Theory of Sound* for a generalisation of this work to other geometries, using other solutions of the wave-equation. (The basic physics is nearly all contained in the one-dimensional problem).

We consider a large mass of liquid, bounded by a solid plane wall at $x = 0$ and extending to an infinite distance to the right. We now suppose the wall set into sinusoidal oscillations in a direction perpendicular to its plane, so that at a time t its position is $x = x_0 \cos(kct)$. We shall neglect small quantities of the order of x_0^2. We now look for a velocity potential that shall satisfy two conditions:

(a) The x component v_x of velocity must be consistent with the above motion of the wall. That is to say that at $x = 0$

$$-\frac{\partial \phi}{\partial x} = v_x = \frac{dx}{dt} = -kc\, x_0 \sin kct \qquad (6.36)$$

The difference between $\partial \phi / \partial x$ at $x = 0$ and at a distance of the order x_0 from the origin, is of the order of magnitude $x_0 \dfrac{\partial^2 \phi}{\partial x^2}$. By (6.37) below this is of second order in x_0^2, so (6.36) is correct to the first order in x_0.

(b) The velocity potential ϕ must be a solution of the wave-equation and it must describe a sound-wave moving entirely towards the right. In the absence of a reflecting boundary this is required by the physics of the problem. If we had a sound-wave travelling towards the left, it would be setting the solid wall into motion and we should be studying the *inverse* of the sound-creation problem, a situation in which the energy in the sound-wave was being converted into motion of an ear-drum or of a solid body, and so being *absorbed*.

The following velocity potential satisfies both (a) and (b).

$$\phi = cx_0 \cos[k(x - ct)] \qquad (6.37)$$

We have
$$-\frac{\partial \phi}{\partial x} = kcx_0 \sin[k(x - ct)] \qquad (6.38)$$

which agrees with (6.36) if we put $x = 0$. Furthermore, since (6.37) is a function of $(x - ct)$ only, it automatically satisfies the wave-equation and corresponds to a disturbance propagated wholly to the right. From Bernoulli's equation, we can calculate from (6.37) the oscillations of pressure

$$P = P_0 + \rho \frac{\partial \phi}{\partial t} = P_0 + \rho k c^2 x_0 \sin[k(x - ct)] . \qquad (6.39)$$

Equations (6.38) and (6.39) specify the pressure and particle velocity at all times and at all places in the medium, and therefore give a complete description of the disturbance. The velocity potential ϕ itself does not enter into this description; we are, as usual, only concerned with its space and time derivatives.

It is of interest to calculate the rate at which energy is fed into the medium. At any instant the solid boundary is moving forwards or backwards and the rate at which it does work per unit area is the pressure at the portion of medium in immediate contact with the boundary (given sufficiently accurately by putting $x = 0$ in (6.39)) multiplied by the rate, $v_x = -kcx_0 \sin(kct)$, at which the solid boundary is moving forwards. That is to say, we have for the rate of doing work

$$P kcx_0 \sin(kct) = P_0 kcx_0 \sin(kct) + \rho k^2 c^3 x_0^2 \sin^2(kct) . \qquad (6.40)$$

Sometimes the boundary is imparting energy *to* the medium, sometimes work is being done on it *by* the medium. In (6.40) the first term is as often positive as negative, while the second term is always positive (except that it is zero at two instants during each oscillation). This means that in the long run work is being done by the wall *on* the medium. Over a large number of periods, the average value of $\sin^2(kct)$ is $\frac{1}{2}$, so we conclude from (6.40) that energy radiated from the wall is transferred to the medium at an average rate $\frac{1}{2}\rho k^2 c^3 x_0^2$ per unit area. It is possible to verify that this energy is continually being transmitted through the medium with the velocity c. We can use (6.38) to calculate the kinetic energy at any point in the medium, and from the pressure (6.39) and the equation of state we can calculate the potential energy for all values of x and t. Each of these is found on average to be $\frac{1}{4}\rho k^2 c^2 x_0^2$ per unit area, which means that energy is being propagated through the medium with a velocity c. It is a foregone conclusion that the equation of energy balance must be satisfied, because the equations of hydrodynamics were derived from mechanics and an energy-equation always exists for conservative mechanical systems. As in the mechanics of small oscillations, the largest surviving terms in the energy are small quantities of the *second* order.

If we use an expression of type (6.35) instead of one of (6.33) for ϕ, it is easy to modify this theory to describe the creation of sound by a spherically symmetrical source. Other types of source are dealt with in Rayleigh's *Theory of Sound*.

6.7.2 Transmission of sound across a boundary between two media

We have implicitly assumed that the material of the solid boundary exciting the sound is incompressible (or of very small compressibility compared with the liquid or gaseous medium.) It is possible to generalise the above work to allow for a finite compressibility of the material of the wall. The relative motions of different parts of the wall are then governed by the equations of elasticity, and the problem becomes very similar to the problem of the **transmission** of sound from one medium to another. We shall now discuss this problem.

We again restrict ourselves to plane waves. We suppose that the plane $x = 0$ is the boundary between a medium of density ρ_1 and velocity of sound c_1 on the left, and another medium of density ρ_2 and velocity of sound c_2. We shall assume a plane wave incident normally on the boundary from the left. We shall show that (Fig. 6.7) all conditions can be satisfied if we postulate both a wave reflected back into the first medium and a wave transmitted into the

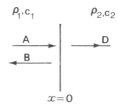

Fig. 6.7 Reflection and transmission of a plane boundary.

second medium and propagated wholly to the right. On physical grounds of course we would not expect any wave propagating to the left in the second medium unless there were another reflecting boundary. We therefore assume the following velocity potentials.

In medium 1 $\phi = \underset{\text{(Incident)}}{A \cos [k_1(x - c_1 t)]} + \underset{\text{(reflected)}}{B \cos[k_1(x + c_1 t)]}$.   (6.41)

In medium 2 $\phi = \underset{\text{(Transmitted)}}{D \cos[k_2(x - c_2 t)]}$.   (6.42)

Our boundary conditions are that the normal velocity and the pressure are both continuous across the boundary, that is to say that we must have $\partial \phi/\partial x$ and $\rho \dfrac{\partial \phi}{\partial t}$ the same on both sides, putting $x = 0$ after the differentiations. Hence

$$\frac{\partial \phi}{\partial x}: \quad k_1(A - B)\sin(k_1 c_1 t) = Dk_2 \sin(k_2 c_2 t) \qquad (6.43)$$

and $\rho \dfrac{\partial \phi}{\partial t}$: $\rho_1 k_1 c_1 (A + B) \sin(k_1 c_1 t) = \rho_2 k_2 c_2 D \sin(k_2 c_2 t)$.   (6.44)

These boundary conditions must hold at all times and not just for one particular value of t, which is possible if $k_1 c_1 = k_2 c_2$. This frequency condition, that the frequency of the oscillation is the same in both media, might have been anticipated on physical grounds. We then have

$$A - B = \frac{D k_2}{k_1} = \frac{D c_1}{c_2} \text{ (by the frequency condition)} \qquad (6.45)$$

$$A + B = \frac{\rho_2 k_2 c_2 D}{\rho_1 k_1 c_1} = \frac{\rho_2 D}{\rho_1}, \qquad (6.46)$$

which enable us to calculate B and D if A is given. However, because of the linearity of the wave-equation we only get the *ratios* B/A and D/A, which we may call reflection and transmission coefficients. These are ratios of amplitudes; if we want ratios of energies transmitted and reflected we must square these quantities. We notice that, according to these equations, $B = 0$ if

$$\rho_1 c_1 = \rho_2 c_2 . \qquad (6.47)$$

That is to say that there is no reflection at the boundary if this condition is satisfied, all the energy being transmitted into the second medium. The condition for this is known as **acoustic matching**. It is of considerable technical importance in two quite different contexts:

(i) If we are designing ultrasonic equipment we want to transmit as much energy into the medium as possible, so we try to approach the condition (6.47). For a solid-liquid or solid-gas boundary this can be difficult, because both ρ and c are usually larger for the solid. As in optics, one can sometimes improve the transmission efficiency by introducing an intermediate layer of material so that the change in ρc takes place in two smaller steps instead of one larger one.

(ii) We may be trying to reduce the penetration of sound into a certain region. One way of doing this is by interposing one or more reflecting boundaries with a large jump in the quantity ρc. So-called 'sound absorbing' materials like foam rubber do not significantly absorb sound in the sense of converting it into heat, but they do act rather like a material with alternate layers of solid and air. In addition, it is shown in Rayleigh's *Theory of Sound* that a sphere of gas in a liquid or solid **scatters** sound in all directions. This effect is also increased if we increase the difference in ρc. Both ρ and c are much less for a gas than they are for a liquid or solid, so we expect a medium like foam rubber to be a highly efficient **reflector** and **scatterer** of sound. In underwater acoustics the scattering effect of minute gas bubbles in the sea is of major importance.

It is quite easy to generalise the above theory of reflection of plane waves to the case of oblique incidence. For a plane wave incident on the boundary at an angle θ, we replace (6.41) and (6.42) by

Medium 1: $\phi = A \cos[k_1(x\cos\theta + y\sin\theta - c_1 t)] +$
$$+ B \cos[k_1(-x\cos\theta' + y\sin\theta' - c_1 t)] \qquad (6.48)$$

Medium 2: $\phi = D \cos[k_2(x\cos\psi + y\sin\psi - c_2 t)]$.   (6.49)

These quantities are solutions of the wave equation for all values of the angles θ, θ' and ψ. We obtain (6.48) and (6.49) simply by *rotating the coordinate x and y axes,* starting with the solutions (6.41) and (6.42) that describe waves propagated along the x axis. We now show that all the boundary conditions are satisfied provided that these angles obey the same laws of reflection and refraction as in optics. As before, we impose the condition that $\partial\phi/\partial x$ and $\rho\dfrac{\partial\phi}{\partial t}$ are continuous at the boundary, given by $x = 0$.

Because the boundary conditions must hold for all values of the time, we obtain the frequency condition $k_1 c_1 = k_2 c_2$ as before. The laws of reflection and refraction follow because the boundary conditions must hold for *all* values of y on the plane $x = 0$. From this we conclude that

$$\sin\theta = \sin\theta' \quad \text{(Law of reflection)}. \qquad (6.50)$$

Also $k_1 \sin\theta = k_2 \sin\psi$ or $\dfrac{\sin\theta}{c_1} = \dfrac{\sin\psi}{c_2}$ (Snell's Law of refraction)(6.51)

(6.45) is replaced by

$$k_1(A - B)\cos\theta = k_2 D \cos\psi , \qquad (6.52)$$

while (6.46) is unchanged

$$\rho_1 k_1 c_1 (A + B) = \rho_2 k_2 c_2 D . \qquad (6.46)$$

Strictly, this means that acoustic matching can only hold for one angle of incidence. In practice, the factor $\cos\psi/\cos\theta$ by which (6.52) differs from (6.45) only varies very slowly with θ, so the point is of little importance except at high angles of incidence.

Sound refraction does indeed occur, and is of great importance in underwater acoustics because the velocity of propagation is sensitive to temperature and salinity. 'Total internal reflection' of sound occurs when $c_2 \sin\theta > c_1$, just as it does in optics, but it can be observed only as a laboratory experiment. If the sea is stratified in temperature and salinity c can vary rapidly with depth, leading to a wide variety of acoustic effects. In particular, we can get effects analogous to those produced by **wave-guides** in electromagnetics, or we can get abnormally rapid decay of amplitude with distance.

We see that there is a close analogy between acoustics and optics, but important differences are due to the fact that pressure, velocity potential and condensation are essentially **scalar** quantities, while electric and magnetic fields are **vectors**. Sound waves are **longitudinal**, electromagnetic waves in general are **transverse**. The relations (6.46) and (6.52) are the analogues of Fresnel's relations in optics, but their derivation is very much simpler because we are only concerned with a single scalar quantity, whereas in electromagnetics we have to satisfy boundary conditions for both the vector quantities **E** and **H**.

6.8   VISCOUS FLOW

Up till now, we have assumed that liquid layers can slip past one another tangentially without exerting any force. This is obviously unrealistic if the molecules interact with one another. It is important to be able to estimate numerically the effect of neglecting viscosity in a given situation, and a great deal of work has been done on viscous hydrodynamics. In general, it is necessary to drop the assumption that pressure is a scalar quantity and to replace it by a **stress-tensor** similar to the corresponding quantity in elasticity. The reader is referred to larger works such as Lamb's *Hydrodynamics* for details. Some situations, such as the viscous damping of sound or of surface waves, are relatively easy to handle. Others, like flow of a viscous liquid round an obstacle, can only be treated approximately.

The following situations are important because they occur in instruments that actually measure viscosity.

6.8.1 Flow at constant rate of shear

We define viscosity as tangential force per unit area per unit velocity gradient. We consider in Fig. 6.8 a situation not unlike that which occurs in a practical viscometer, in which a flat layer of liquid is confined between two flat plates spaced apart at a distance a. Suppose the bottom plate is fixed and the top

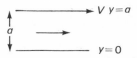

Fig. 6.8 Viscous flow between flat plates.

plate is dragged to the right in the x-direction with a tangential velocity V by a force F. If we assume (which would be almost exactly true for liquids) that there is no slip at either solid surface, it is clear that we can satisfy everything by assuming that $v_x = Vy/a$, that is, the x-component varies linearly with y. For many liquids it is found that the force F which must be applied to the top plate to maintain a constant velocity V is proportional to V. Such liquids are called **Newtonian** liquids. By definition, the force P per unit area exerted on *any* layer of liquid parallel to the x axis is

$$P = \frac{\eta V}{a} \tag{6.53}$$

where η is the viscosity which is measured in Nsm^{-2}.

In particular, this will be the tangential force exerted *on* the liquid by the moving plate and *by* the liquid on the fixed plate. It is easy to modify this treatment to describe the motion of viscous liquid bètween two concentric cylinders. If the outer one is set into axial rotation, the inner one experiences a torque proportional to the viscosity. However, end-effects are hard to calculate properly and such an instrument usually has to be calibrated by using in it a liquid of known viscosity.

6.8.2 Flow in a cylindrical tube

Poiseuille's method of measuring viscosity by determining the rate of flow through a cylindrical tube of radius a at a given pressure-difference between its ends involves a more complicated velocity profile. As before, we assume no slip at the wall, so that $v_x = 0$ at $r = a$. However, for the present we assume $v_x = v(r)$, i.e. that the flow is axial and symmetric. Such flow is sometimes called **laminar flow** in contrast with turbulent flow, which occurs at high pressure differences. We consider the forces on a hollow cylinder of liquid of axial length dx and of inner and outer radii r and $r + dr$. The force on the outer surface is, by the definition of viscosity,

$$\eta \, 2\pi(r + dr) \, dx \, \frac{\partial v}{\partial r}. \tag{6.54}$$

This force is tending to hold the element back, since v decreases as r increases. The force on the inner surface of the element is tending to pull it forward. The net decelerating force on the element has to be equated to the force due to the pressure-differences along the length of the tube, $= -\dfrac{\partial P}{\partial x} \times dx \times$ area of end of hollow cylinder. Thus we have

$$\frac{\partial P}{\partial x} 2\pi r \, dr \, dx \;=\; 2\pi \eta \, dx \left[\left(r \frac{\partial v}{\partial r} \right)_{r+dr} - \left(r \frac{\partial v}{\partial r} \right)_{r} \right].$$ (6.55)

In calculating the viscous forces on the element we have had to allow for the differences in area of the inner and outer surfaces, as well as for a possible change in the velocity gradient $\partial v/\partial r$. If we make the natural assumption that at any value of x the pressure is the same over the whole cross-section of the tube, we have from (6.55) that $\dfrac{1}{r} \dfrac{\partial}{\partial r} \left(r \dfrac{\partial v}{\partial r} \right)$ must also be independent of r. Let it be equal to A. Then integrating we get $r \dfrac{\partial v}{\partial r} = \dfrac{Ar^2}{2}$. (The constant of inte-gration must be zero because otherwise $\partial v/\partial r$ would be infinite at $r = 0$.) Integrating again, we have $v = \dfrac{Ar^2}{4} + B$. We can determine A and B in terms of v_0, the velocity at $r = 0$, from our boundary condition that the velocity must vanish at $r = a$. We have finally

$$v \;=\; \frac{v_0(a^2 - r^2)}{a^2}.$$ (6.56)

In other words, the velocity profile is a parabola. Inserting this expression into (6.55) we find

$$-\frac{\partial P}{\partial x} \;=\; \frac{4\eta v_0}{a^2}.$$ (6.57)

In practice, v_0 is difficult to measure directly with any accuracy, and it is usually inferred indirectly by measuring the total volume of fluid flowing through the tube in unit time. This is $Q = \displaystyle\int_0^a 2\pi r \, dr \, v$, which on inserting v from (6.56) is equal to $\dfrac{\pi}{2} v_0 a^2$. Thus, we obtain finally

$$Q \;=\; -\frac{\pi a^4}{8\eta} \frac{\partial P}{\partial x}$$ (6.58)

for the total rate of discharge through a tube under a steady pressure gradient. This is known as **Poiseuille's law** and is the basis of a very accurate method of measuring viscosity.

In situations of slow viscous flow, i.e. when the flow is mainly controlled by viscosity, the assumption of irrotational motion cannot be made. Indeed it would be false in either of the simple situations just discussed.

REFERENCES

Lamb, H., *Hydrodynamics, 6th Edition,* 1932, Cambridge University Press.

Milne-Thomson, L. M., *Theoretical Hydrodynamics, 5th Edition,* 1968, Macmillan, London.

Rayleigh, Lord, *Theory of Sound, Vols I and II*, 1945, Dover reprint.

Tabor, D., *Gases, Liquids and Solids,* 1969, pp. 62–64, Penguin.

Chapter 7
ULTRASONIC WAVES IN LIQUIDS

7.1 INTRODUCTION

A sinusoidal sound wave propagated through a liquid is an elastic wave and is accompanied by periodic changes of pressure. The frequency of such waves can vary between a few hertz and several gigahertz (1 GHz = 10^9Hz). Frequencies up to about 15 kHz lie within the audible range while those between 15 kHz and about 800 MHz are said to be in the ultrasonic range; frequencies above 10^9 Hz lie in the so-called 'hypersonic' range.

In the present chapter we shall be concerned with the propagation in liquids of sound waves with frequencies mainly in the ultrasonic range. After an initial summary of wave propagation based on the theory of the previous chapter, followed by a discussion of the attenuation of sound waves in a liquid, there is a general account of relaxation effects. The attenuation and relaxation effects caused by viscosity, heat conduction and molecular phenomena are then considered. This is followed by an account of the various experimental techniques which have been employed, and a brief discussion of hypersonic waves.

7.2 A GENERAL DISCUSSION OF WAVE PROPAGATION IN A FLUID

An ultrasonic wave is propagated as a longitudinal wave in a liquid or gas because the particles of the fluid oscillate in a direction which coincides with that of the wave. This is because a fluid medium possesses only one elastic modulus, the bulk modulus K. The propagation velocity, c, of the wave depends on this modulus and on the density ρ_0 of the fluid according to the equation, equivalent to (6.3),

$$c = \sqrt{(K_s/\rho_0)}. \tag{7.1}$$

The suffix s emphasizes the fact that we must use the adiabatic bulk modulus, K_s. Equation (7.1) holds for small excess pressure amplitudes for which ρ_0 can be considered to remain a constant. This **excess** or **acoustic** pressure, p, is equal to the difference between the instantaneous pressure P and the ambient (hydrostatic) pressure P_0, that is $p = P - P_0$.

Let us consider a plane wave travelling in a liquid in the x-direction. Its propagation is given by the well known wave equation (cf. Chapter 6.7)

$$\frac{\partial^2 p}{\partial t^2} = c^2 \frac{\partial^2 p}{\partial x^2}. \tag{7.2}$$

As the wave proceeds, each particle of the liquid suffers a certain displacement from its mean position, we denote this **particle displacement** by ξ. The corresponding **particle velocity** u in the medium is given by $u = d\xi/dt$ and the **particle acceleration** a by $a = du/dt = d^2\xi/dt^2$.

Another useful quantity is the **velocity potential** ϕ introduced in Chapter 6.5. This is related to the three components v_x, v_y, v_z of the particle velocity in the three-demensional case by the equations $v_x = -\partial\phi/\partial x$, $v_y = -\partial\phi/\partial y$ and $v_z = -\partial\phi/\partial z$.

In equation (7.2) the propagation of the sound wave has been described in terms of the acoustic pressure, p; in Chapter 6 it was shown that this equation is also satisfied by a number of other quantities such as ϕ, ξ and $\dot{\xi}$. For example, in terms of the particle displacement ξ the equation would be

$$\frac{\partial^2 \xi}{\partial t^2} = c^2 \frac{\partial^2 \xi}{\partial x^2}. \tag{7.3}$$

As in Chapter 6, we consider the typical solution

$$\xi = \xi_0 \exp j(\omega t - kx) \tag{7.4}$$

describing waves travelling in the positive x-direction. In this equation ξ_0 represents the peak value of the particle displacement, $\omega = 2\pi f$ is the pulsatance (angular frequency) and $k = \omega/c = 2\pi/\lambda$ is the wave number.

Another useful concept is the condensation s, defined in Chapter 6.7 as the fractional change of density from the mean value, that is $s = (\rho - \rho_0)/\rho_0$. Except in the case of shear waves s is the same as the *strain* S given by $S = -\partial\xi/\partial x$. Two further results, which we shall use later, are

$$p = \rho_0 c^2 S \tag{7.5}$$

and
$$\frac{\partial p}{\partial x} = -\rho_0 \frac{\partial^2 \xi}{\partial t^2}. \tag{7.6}$$

The velocity of sound in liquids varies from about 900 to 2000 m/sec. For example, at 20°C the values of c for carbon tetrachloride, water and pure glycerol are respectively 950, 1490 and 1940 m/s. The value of c also depends on the temperature and pressure. For most pure liquids at temperatures far from the critical values the value of c decreases as the temperature increases. Water is an exception to this general rule, and in this case c increases by about 2m/s for each degree Celsius rise in temperature and reaches a maximum at 73°C; above this temperature c decreases as the temperature rises further. As far as pressure changes are concerned, the velocity c increases in an approximately linear way with pressure for all liquids. This is simply a consequence of the fact that the bulk modulus increases as the molecules are squeezed closer together. For water c increases by about 0.2m/s for every pressure increase of one atmosphere.

7.3 ATTENUATION OF ULTRASONIC WAVES IN LIQUIDS

In most cases the intensity of a sound wave decreases continuously as it is propagated through a liquid. This is generally referred to as propagation loss. This propagation loss consists in general of two contributions, namely spreading loss and attenuation loss. Spreading loss occurs in waves diverging as they progress and is the loss of intensity caused by the geometrical spreading of the waves. For example, the intensity of spherical waves would fall off according to the inverse square law. Spreading loss does not depend on the nature of the liquid. On the other hand attenuation loss is characteristic of the particular liquid and is due to the combined effects of scattering and absorption within the medium. In this section we consider this attenuation loss only.

If we consider a plane wave at a given instant of time t, the attenuation loss may be described by the equation

$$I_2 = I_1 e^{-2\alpha(x_2 - x_1)} \tag{7.7}$$

where I_1, I_2 are the respective intensities at $x = x_1$ and $x = x_2$ and α is the absorption coefficient. Equation (7.7) implies that the fractional decrease of intensity per unit distance is constant, which is what is usually found in practice, and α is measured in neper/m. The fall-off of excess pressure corresponding to equation (7.7) would be

$$p_2 = p_1 e^{-\alpha(x_2 - x_1)} \tag{7.8}$$

since the intensity is proportional to p^2.

It is often better to express the attenuation loss in decibels; this can be done as follows. From equation (7.7) we have

$$\log_{10}(I_1/I_2) = 2\alpha(x_2 - x_1)/2.303$$

so that $10 \log_{10}(I_1/I_2) = 20\alpha(x_2 - x_1)/2.303$

or $10 \log_{10}(I_1/I_2) = \alpha^1(x_2 - x_1)$.   (7.9)

The quantity $10 \log_{10}(I_1/I_2)$ is the attenuation loss measured in decibels and α^1 on the right-hand side of this equation is the absorption coefficient expressed in decibels per metre.

When absorption is taken into account equation (7.4) becomes

$$\xi = \xi_0 \exp(-\alpha x) \exp j(\omega t - kx)$$ (7.10)

Sound absorption in a liquid, which ultimately results in all the energy of the wave being converted into heat by the non-ideal nature of the liquid, is due to three main causes: viscosity, thermal conduction and relaxation effects. These are considered in greater detail in Chapter 7.5 and 7.6. It should be realised however that sound absorption can be discussed phenomenologically as being due to a time lag between the pressure variation in a sound wave and the associated density variation. The three above effects can be considered from this unified standpoint; such a treatment is given in *Sonics* by Hueter and Bolt (1955).

7.4   RELAXATION EFFECTS: A GENERAL DISCUSSION

We have already mentioned in Chapter 7.3 that the attenuation of a *plane* wave in a liquid is caused by scattering and absorption. In scattering, some of the energy of the initially parallel beam is deflected into other directions; the attenuation thus caused can enable an estimate of the size of the sources of scattering (for example, the gas bubbles in a liquid) to be made. Absorption losses are caused by certain effects characteristic of the liquid; these are called **relaxation effects**. We will devote this section to a general discussion of these effects and then proceed later to the causes of such effects in liquids.

It is helpful to begin with an electrical analogy and to consider the circuit shown in Fig. 7.1, where Z is a complex impedance. If a key is closed we apply a step-function voltage V_0 to the circuit. Consider now the two following cases.

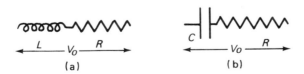

Fig. 7.1 Voltage applied across a complex impedance Z.

(i) If Z consists of a resistance R and inductance L in series, then the current I takes a finite time to reach its final steady-state value I_0. We can say that the current 'relaxes' from its initial to its final value; the changes in I do not follow, but lag behind, those in V.

(ii) If Z consists of a resistance R and capacitance C in series, the charge Q across C also takes a finite time to relax to its final steady value Q_0.

The behaviour of both (i) and (ii) can be expressed mathematically in identical ways. Let us denote by A the quantity which lags behind the cause (in this case, the applied voltage) producing it, and let A_0 be the final equilibrium value of A. In case (i) $A = I$ and in case (ii) $A = Q$. Then the well known result for both circuits (i) and (ii) may be written as

$$A = A_0(1 - e^{-t/\tau}) \tag{7.11}$$

and the rate dA/dt at which A tends to A_0 is proportional to $(A_0 - A)$ for, on differentiating, we have

$$\frac{dA}{dt} = \frac{1}{\tau}(A_0 - A). \tag{7.12}$$

In general a process described by (7.12) is called a **relaxation process** and the rate at which the dependent variable A relaxes to its final value A_0 is determined by τ, the **relaxation time**. τ is defined as the time taken for A to increase from zero to the value $(1 - 1/e)A_0$.

Relaxation phenomena can arise in a liquid by the propagation of low-amplitude plane waves in the liquid. Following Blitz (1967) we shall compare the acoustic case with that of the R-C circuit considered above. The charging of the capacitor C through a resistor R is governed by

$$V = Q/C + R \, dQ/dt. \tag{7.13}$$

Blitz shows that there is a similar relation between the acoustic pressure p and the strain $S = -\partial\xi/\partial x$ when a small amplitude plane wave is propagated through a liquid and writes, by analogy with equation (7.13).

$$p = \rho_0 c^2 S + r \, dS/dt. \tag{7.14}$$

The quantities V, Q, $1/C$ and R correspond respectively to p, S, $\rho_0 c^2$ and r; r is thus to be regarded as 'resistance', that is, some frictional constant. S lags behind p and relaxes to its final steady value $S_0 = p_0/\rho_0 c^2$ (where p_0 is the final steady value of p) with a relaxation time τ given by

$$\tau = r/\rho_0 c^2. \tag{7.15}$$

If we substitute $S = -\partial\xi/\partial x$ into (7.14) and differentiate with respect to x we have

$$\frac{\partial p}{\partial x} = -\rho_0 c^2 \frac{\partial^2 \xi}{\partial x^2} - r \frac{\partial^3 \xi}{\partial x^2 \partial t} ,$$

and using equation (7.6) then we have, on substituting and rearranging,

$$\frac{\partial^2 \xi}{\partial t^2} = c^2 \frac{\partial^2 \xi}{\partial x^2} + \frac{r}{\rho_0} \frac{\partial^3 \xi}{\partial x^2 \partial t} . \tag{7.16}$$

We assume that we have an attenuated plane simple harmonic wave given by equation (7.10). Differentiating equation (7.10) we have

$$\frac{\partial^2 \xi}{\partial t^2} = -\omega^2 \xi, \quad \frac{\partial^2 \xi}{\partial x^2} = (\alpha + jk)^2 \xi, \quad \frac{\partial^3 \xi}{\partial x^2 \partial t} = j\omega(\alpha + jk)^2 \xi$$

and hence, on substituting in (7.16) we get

$$-\omega^2 = c^2 (\alpha + jk)^2 + jr\omega(\alpha + jk)^2/\rho_0 . \tag{7.17}$$

Equating the imaginary parts gives us

$$k^2 - \alpha^2 = 2k\alpha c^2 \rho_0/\omega r$$

that is, $k^2 - \alpha^2 = 2k\alpha/\omega\tau.$ \hfill (7.18)

Equating the real parts and using (7.18) gives us

$$\alpha = \omega^3 \tau/2kc^2 (1 + \omega^2 \tau^2) . \tag{7.19}$$

At very low frequencies the various changes induced by the pressure variations in the wave will be able to 'follow' these slowly-varying pressure changes, that is, relaxation effects are absent or at least negligible. We shall take c as being the velocity of sound under these conditions. At higher frequencies $f = \omega/2\pi$, where such relaxation effects are not negligible, we let $v = \omega/k$ represent the velocity; at such frequencies v varies with frequency and this variation is called **velocity dispersion**. Substituting $k = \omega/v$ into (7.19) we get

$$\alpha = \omega^2 \tau v/2c^2 (1 + \omega^2 \tau^2) \tag{7.20}$$

showing that the attenuation also varies with frequency.

The amplitude loss per cycle, that is the logarithmic decrement δ, is given by

$$\delta = \frac{2\pi\alpha}{k} = \frac{\pi v^2 \omega \tau}{c^2(1 + \omega^2\tau^2)}$$

If the velocity dispersion is small then v^2/c^2 may be regarded as being constant and

$$\delta = \text{const.} \times \frac{\omega\tau}{(1 + \omega^2\tau^2)}$$

showing that δ, and therefore α, is a maximum when $\omega\tau = \omega_0\tau = 1$. This value of ω corresponds to maximum absorption, and the corresponding **relaxation frequency** f_0 is given by

$$f_0 = \omega_0/2\pi = 1/2\pi\tau . \tag{7.21}$$

For frequencies $f \ll f_0$, $\omega^2\tau^2 \ll 1$, then from (7.20)

$$\alpha \simeq \omega^2\tau v/2c^2$$

$$= \omega^2\tau/2c \tag{7.22}$$

very nearly since $v \simeq c$ at these low frequencies. So α is proportional to ω^2 (or to f^2) under these conditions.

As already mentioned, there are three main causes for absorption and velocity dispersion in liquids. These are due to viscosity, thermal conduction and molecular phenomena. The first two are predicted on classical grounds and their combined effect is called 'classical' absorption. These three causes of absorption are all of the relaxational type, and we now proceed to consider each in turn. Before doing this it should be mentioned that the above ideas are very similar to the work of Debye on dielectric relaxation accompanying an applied alternating electric field. Much of the mathematics is the same as well. For further details see Chapter 5 of *Molecular Behaviour* by Mansel Davies.

7.5   LOSSES DUE TO VISCOSITY AND HEAT CONDUCTION

We now consider the viscous attenuation of a plane wave in a liquid of viscosity η. As the wave travels there is a relative motion of adjacent layers of the liquid and this results in the creation of viscous forces which act against the acoustic pressure due to the wave. Energy is taken out of the wave to overcome these viscous forces, a progressive extraction of energy which results in a corresponding progressive decrease in the intensity (and hence of pressure, particle displacement, etc.) of the wave. Stokes showed that these effects could be taken into account by modifying (7.3) and writing

$$\frac{\partial^2 \xi}{\partial t^2} = c^2 \frac{\partial^2 \xi}{\partial x^2} + \frac{4\nu}{3} \frac{\partial^3 \xi}{\partial x^2 \partial t}. \tag{7.23}$$

In this equation $\nu = \eta/\rho_0$ is called the kinematic viscosity and is measured in m^2/s.

This equation is of the same form as (7.16) and we therefore conclude that viscous effects are relaxational in nature. The quantity $4\nu/3$ replaces r/ρ_0 and the relaxation time τ_1 is now given by

$$\tau_1 = 4\nu/3c^2 \tag{7.24}$$

(cf. equation (7.15)). When $\omega\tau \ll 1$, which is usually the case, the absorption coefficient α_1 due to viscosity is

$$\alpha_1 = \omega^2 \tau_1/2c = 2\omega^2 \nu/3c^3 \tag{7.25}$$

(cf. (7.22)).

Next, let us consider the role of heat conduction. In adiabatic wave propagation it is assumed that the pressure variations occur so rapidly that the temperature changes in a certain small region without any flow of heat to a neighbouring region. In practice however, a certain amount of heat does flow from a region of high temperature (compression) to an adjacent one of low temperature (rarefaction) and this conduction of heat results in a certain loss of sound energy per cycle. Vigoureux (1950) shows that this absorption due to heat conduction can be accounted for by writing the attenuated plane wave equation as

$$\frac{\partial^2 \xi}{\partial t^2} = c^2 \frac{\partial^2 \xi}{\partial x^2} + \frac{\kappa}{\rho_0 C_v} \left(\frac{\gamma - 1}{\gamma}\right) \frac{\partial^3 \xi}{\partial x^2 \partial t} \tag{7.26}$$

where κ is the thermal conductivity of the liquid, C_v the specific heat at constant volume and γ the ratio of the specific heats. Equation (7.26) is of the same form as (7.16) and again we conclude that this cause of absorption is relaxational.

By an argument similar to that for the viscous case we readily see that the relaxation time is

$$\tau_2 = \frac{\kappa}{\rho_0 c^2 C_v} \left(\frac{\gamma - 1}{\gamma} \right) \tag{7.27}$$

and that the absorption coefficient α_2 due to heat conduction is, assuming $\omega \tau_2 \ll 1$, given by

$$\alpha_2 = \omega^2 \tau_2 / 2c$$

$$= \frac{\kappa}{2\rho_0 C_v} \left(\frac{\gamma - 1}{\gamma} \right) \frac{\omega^2}{c^3} . \tag{7.28}$$

As stated earlier, the absorption coefficients α_1 and α_2 are predicted on classical grounds. Their combined effect is represented by the classical absorption coefficient α_c given by

$$\alpha_c = \alpha_1 + \alpha_2 \tag{7.29}$$

and α_c is also proportional to f^2.

For most liquids $\alpha_2 \ll \alpha_1$, that is, the attenuation due to thermal conduction is negligible compared with that due to viscosity. An exception to this general rule occurs in the case of liquid metals, where thermal conduction makes a greater relative contribution. For fresh water the values are

$$\alpha_1 = 8.1 \times 10^{-15} f^2$$

$$\alpha_2 = 3.0 \times 10^{-18} f^2$$

the units of α_1 and α_2 being neper m^{-1}. Some data for other liquids are given in Table 7.1.

In monatomic liquids such as liquid argon and mercury, the theoretical values of the classical absorption coefficient α_c predicted by the combined effects of viscosity and thermal conduction agree closely with the values measured experimentally. In polyatomic liquids however, the measured absorption coefficient is often very much greater than the classical value predicted by viscosity and thermal conduction. This extra absorption is referred to as **excess** or **anomalous** absorption. The existence of this excess absorption is a very real and important factor because it shows that the description of a liquid based on compressibility, viscosity and thermal conduction is not adequate even on a macroscopic basis. The cause of this excess absorption in polyatomic liquids is known as **molecular relaxation**. In general this is due to two causes known as **thermal relaxation** and also, in associated liquids, **structural relaxation**.

Table 7.1 Calculated values of acoustic velocities and viscous relaxation properties for liquids at 20°C [taken from *Fundamentals of Ultrasonics*, by J. Blitz]

Liquid	Velocity, c (m s^{-1})	τ_1 (s)	α_1/f^2 (neper m^{-1}s^2)
Acetone	1200	3.8×10^{-13}	6.3×10^{-15}
Benzene	1320	5.6×10^{-13}	8.4×10^{-15}
Carbon Tetrachloride	950	9.2×10^{-13}	2.0×10^{-14}
Castor Oil	1500	6.2×10^{-10}	8.1×10^{-12}
Chlorobenzene	1320	6.8×10^{-13}	1.0×10^{-14}
Ethyl alcohol	1200	1.7×10^{-12}	2.9×10^{-14}
Glycerol (pure)	1940	4.2×10^{-10}	5.7×10^{-12}
Methyl alcohol	1120	7.9×10^{-13}	1.4×10^{-14}
Nitrobenzene	1480	1.1×10^{-12}	1.4×10^{-14}
Olive Oil	1440	6.0×10^{-11}	8.2×10^{-13}
Toluene	1320	5.2×10^{-13}	7.7×10^{-15}

The concept of an 'associated liquid' is somewhat vague but has appeared repeatedly in the literature. We use it here to describe the behaviour of liquids whose intermolecular interaction is too complicated to be described by simple potential functions of the kind discussed in Chapter 3. Such interactions occur, for example, between molecules that have one or more hydroxyl or amino groups. The particular quantum mechanical effect known as 'hydrogen bonding' has the effect of producing a very localised interaction between molecules containing such groups, which is negligible unless two such groups of different molecules are near one another. If they are close to one another the two molecules are loosely linked, and we can have an appreciable gain of energy which is quite distinct from the London-van der Waals attraction described in Chapter 3. If there is only one such group per molecule, as in the ordinary alcohols, the molecules are loosely bound together in pairs. If there are two or more, as in glycerol, there is the further possibility of forming rings and short chains. Since the forces are fairly weak, the pairs or complexes of molecules can have characteristic frequencies of oscillation in the ultrasonic range. See Chapter 7.6.2 below.

In Table 7.2 a classification of liquids based on their ultrasonic properties is given. Liquids in groups 1, 2 and 3 exhibit excess absorption, whilst those in groups 4 and 5 have normal or near-normal (that is, classical) values for their absorption coefficients.

Table 7.2 Acoustic classification of liquids.

Group	α/α_c	$d\alpha/dT$	Type of liquid	Examples
1	3–1500	positive	Unassociated polyatomic	CS_2, $CC\ell_4$, C_6H_6
2	1.5–3	negative	Associated polyatomic	Water, alcohols
3	5–5000	positive or negative	Organic acids, esters	CH_3COOH, $CH_3COOC_2H_5$
4	1	positive	Monatomic, diatomic	Hg, liquid A, liquid He, liquid O_2, H_2
5	About 1	negative	Associated polyatomic	Glycerin, castor oil, highly viscous liquids

7.6 MOLECULAR RELAXATION

7.6.1 Thermal relaxation

Every polyatomic molecule possesses a number of degrees of freedom; these are of two kinds, external and internal. The external degrees of freedom (also possessed by a monatomic molecule) are three in number corresponding to the three translational velocity components of the molecule. This translational energy is the one which is directly obtained from the acoustic wave, and it is passed on from molecule to molecule by collisions with virtually no time-delay. The internal degrees of freedom correspond to the **vibrational** and **rotational** motion of the molecules. One of these internal modes can be excited if the necessary energy can be transferred from one of the external degrees of freedom. Such a transfer of energy from an external to an internal degree of freedom requires a certain finite relaxation time, so that the internal modes lag slightly behind the external modes. This is called thermal relaxation and the detailed theoretical treatment involves a consideration of the transition from a translational energy state to an excited internal state. The reader interested in such a detailed treatment is referred, for example, to Blitz (Chapter 4) or Gooberman (Chapter 6). It is found that the absorption coefficient due to thermal relaxation, α_{relax}, is again proportional to f^2.

Two kinds of thermal relaxation not due to the same mechanism have been found in liquids. These are due to **vibrational relaxation** and **rotational isomerism**. They are both referred to as thermal relaxation because both are essentially due to the disturbing influence on a previously existing equilibrium of the temperature changes caused when an ultrasonic wave is propagated through the liquid.

(i) **Vibrational relaxation.** In this case the energy of the translational (external) degrees of freedom is transferred to the three vibrational (internal) modes and the corresponding relaxation times are of the order of 10^{-10} sec or less. The work of Andreae, Heasall and Lamb (1956) on liquid carbon disulphide has revealed this type of thermal relaxation. Examples of other liquids in which this mechanism is present are those in group 1 in Table 7.2.

(ii) **Rotational Isomerism.** Molecules of certain chemical compounds can exist in two different geometrical forms possessing different energies. The two forms are known as isomers and both arise from a rotation of groups about the valence bonds. Examples are acrylaldehyde, ethyl acetate and cyclohexene. In this type of liquid an equilibrium will exist between the two isomers at any particular temperature. When an acoustic wave is propagated, the accompanying temperature changes upset the equilibrium and the relative amounts of the isomers change. That is, a relaxation from one isomeric form to the other occurs with a consequent absorption of acoustic energy. The relaxation times involved here are much longer than those in vibrational relaxation and are of the order of 10^{-6} sec.

For a fuller discussion of thermal relaxation the reader is referred to Lamb (see Mason, p. 203).

7.6.2 Structural relaxation

This occurs in associated polyatomic liquids such as water and the alcohols. It is due to the periodic pressure changes and is not affected by temperature variations. To understand it we invoke the quasi-lattice model (see Section 2.3) in which the liquid is pictured as being a loosely-packed crystalline solid. In a compressional half-cycle the positive excess pressure causes the molecules to be 'squashed together' and this is achieved by molecules moving into empty gaps in the lattice. During a rarefaction half-cycle, the less closely-packed molecular arrangement is restored. Since these changes in the local order of the molecules lag behind the pressure changes, the process is a relaxational one and results in absorption and velocity dispersion. Hall (1948) has investigated this type of absorption in the case of water.

Once again the detailed theory shows that $\alpha_{struct.}$ is proportional to f^2.

Furthermore, the relaxation times for structural relaxation are much shorter than those for thermal relaxation.

Formally we can describe structural relaxation by introducing a **bulk** or **compressional** viscosity, η', in addition to the ordinary viscosity η. This is equivalent to saying that energy is dissipated as heat whenever the liquid expands or contracts in addition to the dissipation accompanying shear. Thus $\alpha_1 = 2\omega^2\eta/3\rho_0 c^3$ as given by equation (7.25) becomes modified to $\alpha_1 = 2\omega^2(\eta + \eta')/3\rho_0 c^3$.

7.7   EXPERIMENTAL METHODS

The important quantities which describe acoustic relaxation in liquids are the velocity and absorption coefficient, the so-called propagation constants. It is desirable that these constants be available for as wide a range of frequencies as possible, from low frequencies of about 10^2 Hz to hypersonic frequencies of about 10^{10} Hz. In what follows the more important methods of measuring these constants will be discussed. Detailed descriptions of the apparatus will not, in general, be given but the emphasis will be placed on the principles underlying the methods. The sound waves are produced or received by a **transducer**, a device which converts acoustical energy to or from other forms of energy (i.e. electrical, thermal or mechanical). The effects mainly exploited are piezo-electricity, using materials such as quartz, and magnetostriction, using various alloys. For a full account of such devices the reader is referred to Blitz (Chapter 3).

In methods to be described it is true to say that, in general, the velocity is determined from the time taken to traverse a known acoustic path length. The absorption coefficient α is determined from the equation (7.8), that is,

$$\alpha = \frac{1}{(x_2 - x_1)} \ln(p_1/p_2).$$

7.7.1 The Acoustic Interferometer

In this method a stationary wave system is set up in the liquid between a reversible transducer S and a movable metallic reflector R (Fig. 7.2). The wave

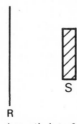

Fig. 7.2 Acoustic interferometer.

system, consisting of nodes and antinodes, will arise when the distance SR is equal to an integral number of half wavelengths of the sound emitted by S. This method is usually used for frequencies in the kilo- and mega-hertz range.

A relatively new method, the acoustic resonator, is a slight variation of the above interferometer. In this case the liquid is contained in a cylindrical cell whose ends consist of two transducers S and R (Fig. 7.3). Sound waves of

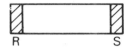

R   S

Fig. 7.3 Acoustic resonator.

wavelength λ are generated in the liquid by exciting S and when $SR = n\lambda/2$ the system will resonate. This resonance frequency is proportional to the sound velocity, and the absorption coefficient can be obtained from the shape of the resonance peak obtained from plotting the output voltage of R against frequency. The acoustic resonator has been used in the range 10^5 Hz to 5×10^7 Hz.

7.7.2 The Pulse Method

In this method, a 'pulse' consisting of a short train of sound waves is generated by a transducer S and sent through the liquid to a second receiving transducer R. The initial amplitude of the pulse leaving S is known and the absorption is estimated by making amplitude measurements of the pulse at various points along its subsequent path in the liquid. The speed of the sound is obtained from the acoustic path length and the time taken.

This pulse technique has been developed by Pinkerton (1949) and by Lamb *et al* (1958), and is now the most widely used method for measurements in the range 5–200 MHz. Some work at even higher frequencies up to about 4×10^{10} Hz has also been carried out by replacing the transducers with specially designed quartz resonators.

7.7.3 Optical Diffraction Method

This method makes use of the fact that the refractive index (for light) of a liquid changes periodically as the compressions and rarefactions due to the sound wave pass along. Thus if a parallel beam of ultrasonic waves is passed through a liquid it can act as a diffraction grating for a parallel beam of monochromatic light waves passing through the liquid in a direction perpendicular to that of the sound waves. In this case a system of stationary sound waves of wavelength λ is set up between the transducer and reflector and, because of the periodic spatial variation of the refractive index of the liquid, a diffraction grating with a spacing $d = \lambda$ is formed. When such a monochromatic beam of light waves of wavelength λ' is diffracted the result

$$n\lambda' \ = \ \lambda \sin \theta \tag{7.30}$$

holds, where n is the 'order' of the spectrum and is an integer and θ is the angle of diffraction.

The central image of a slit source can be regarded as the zero order spectrum and a measurement of the optical intensity of this spectrum can yield values of the absorption coefficient. Blitz has also used this method to make a comparison of velocities in various liquids and liquid mixtures. It is used in the MHz range because this is the frequency which gives a reasonable value of λ in (7.30) when using light of optical frequency.

7.8   HYPERSONIC WAVES IN A LIQUID

The fundamental idea of Debye's very successful theory of solids was that the motions of the molecules, as a result of thermal agitation, could be represented as a superposition of sound waves. The velocities involved would be the velocities of propagation of compressional and shear waves and the frequencies would be the proper frequencies of vibration of the solid, *treated as an elastic continuum*. Debye also assumed, and his guess was confirmed by later work, that there were no modes of oscillation of a real molecular solid of wavelengths smaller than the mean intermolecular spacing. This idea proved to be extremely sound, subsequent explicit calculations on various crystal lattices producing frequency distributions agreeing with Debye's at the low frequency end and not departing greatly from it at higher frequencies. The theory can be developed further to give a good account of the specific heat and equation of state of a solid.

It is natural to ask whether a similar treatment of liquids is possible, and it turns out that it is, subject to one small difficulty. The velocity of ordinary **compressional** sound waves is defined as usual in terms of the adiabatic compressibility, and the possible frequencies of such waves in, say, a liquid contained in a cubical vessel are easily found. As in the Debye theory of solids we associate with each frequency an energy according to Planck's formula. The difficulty arises because it is uncertain in what circumstances a liquid can propagate **transverse** waves, the motion of the molecules being perpendicular to the direction of the propagation of the waves. Such shear waves certainly exist in solids, but we know that a liquid cannot resist static shear. Thus it certainly cannot propagate shear waves of low frequency, though it remains possible that such high-frequency waves may exist. A simple calculation shows that the frequencies of interest in Debye's theory are indeed high, far higher than those of any waves that can be produced experimentally by piezo-electric methods.

According to Planck's formula, waves of a frequency ν will be appreciably excited if $h\nu$ is small compared with kT. Hence we may expect waves of frequencies up to about 10^{13} Hertz to be appreciably excited. Because of the above difficulty we are not certain about the spectrum of transverse frequencies, but the usual assumption is that the number of modes is zero at $\nu = 0$ and rises rapidly as ν increases, again becoming zero for ν very large.

The Landau theory of liquid helium (Chapter 14) is based essentially on the Debye picture. There are thought to be two types of excitation, **phonons** and **rotons**, and these probably correspond roughly to quanta of compressional and transverse sound respectively, but the details are still uncertain.

For a further discussion of hypersonic waves in liquids the reader is referred to Frenkel, *Kinetic Theory of Liquids*, Oxford (1946). The physical reality of hypersonic waves in ordinary liquids can easily be proved experimentally in at least two ways, by neutron scattering and by light scattering. It is also possible to relate the hypersonic spectrum to the two-molecule distribution function $g(r)$ by an argument similar to that for obtaining the characteristic frequencies of a crystalline lattice. Work has also been done on the interpretation of heat conductivity and viscosity in terms of this 'hydrodynamic' picture. They are attributed to the scattering of phonons by one another, but these matters are beyond the scope of this book.

7.8.1 Brillouin scattering by hypersonic waves

We assume that hypersonic (or ultra high frequency) waves exist in a liquid and that they occur as a result of thermal excitation of the quasi-lattice of the liquid. The frequencies of these waves are in the range 10^9–10^{10} Hz, and they are sometimes called thermal phonons. When these hypersonic waves travel through a liquid they give rise, at any given instant of time, to a periodic spatial variation in the refractive index of the liquid. That is, regions of maximum and minimum refractive index are formed as discussed in Chapter 7.7.3 above. The adjacent regions of maximum refractive index may be pictured as forming adjacent reflecting planes at a distance $d = \lambda$ apart in the liquid, where λ is the wavelength of the hypersonic wave.

When monochromatic light of wavelength λ' from a laser is incident on a liquid in which such hypersonic waves exist, it is scattered by the reflecting planes in much the same way as a beam of X-rays is scattered by reflection at the lattice planes of a crystal. Then Bragg's law,

$$\lambda' = 2\lambda \sin \theta$$

for reflection at planes a distance λ apart, holds. In fact, such a scattering of light waves by a liquid through which sound waves are passing was predicted by Brillouin in 1922.

The scattered light is found to contain a wavelength λ' (as in the X-ray case). It must be remembered, however, that the reflecting planes in the liquid are moving with the sound velocity. This gives rise to a Doppler effect in the reflected light waves, because the light waves are reflected at a moving plane. Thus lines of wavelength $\lambda' \pm d\lambda'$ in the scattered light can also be observed where $\pm d\lambda'$ is the Doppler shift occurring. This scattered light is analysed using a Fabry-Perot interferometer. Theory shows that $d\lambda'$ can be related to the sound velocity and the measured value of $d\lambda'$ therefore enables one to determine this velocity.

7.8.2 Neutron scattering by hypersonic waves

The information given by neutron scattering is always fuller than that given by X-ray scattering, because one measures the velocities as well as the intensity of a neutron beam as a function of scattering angle. This means that we can get much more information about the phonons that are doing the scattering; indeed the neutron scattering processes are equivalent to changes in the frequencies and numbers of the phonons present in a liquid. It follows that we can obtain information not only about the *numbers* of phonons of a given frequency but also about their *lifetimes* and, as stated above, the lifetimes of phonons are related to their scattering of one another. As we have already stated, the coefficients of viscosity and thermal conductivity can be related to these scattering processes.

Since a sound-wave involves the coherent motion of a great many liquid molecules, the scattering of a neutron by a hypersonic wave is an essentially different process physically from the scattering of a neutron by a single molecule. The appearance or disappearance of a phonon, or a change in its frequency or direction of propagation, necessarily involves very small changes in the positions and velocities of a great many molecules. (The energy and momentum of an incident neutron are then changed by amounts settled by the conservation laws).

7.9   OTHER WORK

In conclusion a brief mention must be made of a few other applications of ultrasonics in the field. One of the most familiar effects accompanying the propagation of ultrasonic waves in a liquid is the phenomenon known as cavitation which will be discussed in Chapter 8.4.

Some interesting work has also been done in liquids near their critical points, where structural relaxation is thought to play a major role in the high absorption observed. Near the critical point the fluid is compressed to a liquid during the compressional half-cycles and expands to a vapour during the rarefaction half-cycles. Ultrasonic propagation in liquid mixtures has also been investigated.

In chemistry a vast amount of work on fast chemical reactions has been carried out; for a comprehensive account of such investigations the reader is referred to an article by Wyn-Jones (1969).

Mention must also be made of the use of ultrasonics in echo-sounding and submarine detection. We saw in Chapter 6 that some reflection of sound occurs at the boundary of two media that are not acoustically matched. Very recently this effect has been used in medicine — for example the size and position of an unborn baby can be determined with reasonable accuracy.

REFERENCES

Andreae, J. H., Heasall, E. and Lamb, J., 1956, *Proc. Phys. Soc.* **B69**, 625.

Andreae, J. H., Bass, R. Heasall, E. and Lamb, J., 1958, *Acustica,* **8**, 3.

Frenkel, J., 1946, *Kinetic Theory of Liquids,* (Oxford).

Hall, L., 1948, *Phys. Rev.,* **73**, 775.

Pinkerton, J. M. M., 1949, *Proc. Phys. Soc.,* **B62**, 162.

Pinkerton, J. M. M., 1949, *Proc. Phys. Soc.,* **B62**, 186.

Trevena, D. H., 1969, *Contemp. Phys.,* **10**, 601.

Wyn-Jones, E., 1969, *R.I.C. Reviews,* **2**, 59.

SOURCES FOR FURTHER READING

Blitz, J.,*Fundamentals of Ultrasonics,* Butterworths, 1967.

Davies, M., *Molecular Behaviour,* Ch. 5, Pergamon Press, 1965.

Gooberman, G. L., *Ultrasonics,* Chs. 6 and 7, The English Universities Press Ltd., 1968.

Hueter, T. F. and Bolt, R. H., *Sonics,* Ch. 2 and Appendix, Chapman and Hall Ltd., London, 1955.

Mason, W. P., *Physical Acoustics, Principles and Methods,* Vol. II A, Academic Press, New York, 1965. (Reference here to Lamb).

Richardson, E. G., *Ultrasonic Physics,* Chs. 3, 5 and 8, Elsevier Publishing Co., Amsterdam, 1962.

Temperley, H. N. V., *Properties of Matter,* Ch. 7, University Tutorial Press Ltd., 1963.

Tucker, D. G. and Gazey, B.K., *Applied Underwater Acoustics,* Ch. 3, Pergamon Press, 1966.

Vigoureux, P., *Ultrasonics,* Chapman and Hall Ltd., London, 1950.

Chapter **8**
LIQUIDS UNDER TENSION AND COMPRESSION

8.1 INTRODUCTION TO THE BEHAVIOUR OF LIQUIDS UNDER TENSION

It is not generally appreciated that liquids can, under appropriate conditions, sustain very considerable tensions. This property of liquids is of interest in a number of fields in pure and applied science. For example, the engineer interested in ship propellers and hydraulic machinery of all kinds needs to know something of the behaviour of liquids when the pressure near the solid falls rapidly to negative values. Again, the matter is of interest to the physicist since it is related to the molecular theories of the liquid state. In botany, columns of sap transport water from the roots of trees to the topmost leaves. Since many trees are much taller than 33 feet (the atmospheric pressure in terms of a 'water barometer') there must be a negative pressure in the column which transports the sap. This ability of liquids to withstand tension was predicted at the beginning of the last century by Laplace in his theory of capillarity and it has since been the subject of a good deal of investigation, both theoretical and experimental.

8.2 STATIC APPLICATION OF TENSION TO A LIQUID

8.2.1 Berthelot tube experiments

Some of the earlier work of this type was done during the last century and involved some very careful experimentation. One of the first methods was that used by Berthelot (1850); it has since been developed by other workers and is generally known as the **Berthelot tube method**. A Berthelot tube is a sealed tube roughly cylindrical in shape. At ordinary room temperature it is almost completely filled with the liquid, the remaining volume being occupied by air and the liquid vapour. When the tube and its contents are heated the liquid expands at a greater rate than the Berthelot tube. The air is forced into solution and the liquid fills the tube completely at a certain temperature T_f. When the tube is subsequently allowed to cool, the liquid adheres to the walls of the tube and continues to fill it at temperatures below T_f. Thus as the temperature falls below T_f a gradually increasing tension is built up in the liquid until it eventually ruptures at a lower temperature T_b. At the instant of rupture the sudden release of tension is accompanied by a metallic 'click' and a sudden increase ΔV in the external volume of the Berthelot tube; if ΔV can be measured then the critical or breaking tension just before rupture can be determined.

In his original work, using water in a glass tube, Berthelot found that the volume of the water in the tube had increased by 1 part in 240 just before rupture occurred and he estimated that the tension in the water at the same instant was about 50 atm. About a century later Temperley and Chambers (1946) repeated the original experiments of Berthelot and they concluded that the strength of water in the presence of glass is of the order of 30–50 atm. More recently Rees and Trevena (1964a) studied the behaviour of liquids in Berthelot tubes made of steel, and the values they obtained show that the strength of water in a steel tube varies between 10 and 30 atm.

These Berthelot tube results lead to the conclusion that the nature of the walls of the containing vessel in contact with the liquid is most important. As already mentioned, a water-glass system can withstand tensions of the order of 30–50 atm and a water-steel system lower values of 10–30 atm. This is further supported by other experiments by Temperley (1946, 1947), who found that the introduction of a small steel chip, or turning, into a glass Berthelot tube containing water gave values for the breaking tension which were lower than those obtained for glass tubes containing water only. Furthermore, the results he obtained when a steel chip was present agree with those obtained for steel tubes by Rees and Trevena. All this strongly suggests that what is really measured by a Berthelot tube is the adhesion of the liquid to the surface rather than its true tensile strength. In any measurement the value measured seems to be that of the 'weakest link' in the liquid-solid system; for example, in the experiments mentioned above the water-steel link is weaker than the water-glass link.

8.2.2 The centrifugal method

A centrifugal method of stretching a liquid was devised by Reynolds (1878). He used a glass J-tube containing a liquid, in which the long arm was sealed (so that the liquid was simply under its own vapour pressure) or left open to the atmosphere. The latter type of tube is shown in Fig. 8.1. If such a tube is rotated with angular velocity ω about an axis O perpendicular to the plane containing the axis of the tube, a pressure gradient of $\rho\omega^2 r$ is generated in the outward direction at a distance r from the axis of rotation; thus the pressure decreases as we approach the axis. Since the pressure is always atmospheric at A it follows that, as ω is increased, the pressure at O becomes less than atmospheric. As ω is increased still more, the pressure at O becomes negative; that is tension sets in, and when ω is sufficiently large the liquid ruptures. By observing the maximum possible value of ω which did not cause the water to break, Reynolds calculated that the maximum tension which water could withstand was about 5 atm. However, this result is not a reliable one and reasons for discarding it were brought forward by Temperley and Chambers.

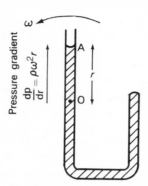

Fig. 8.1 J-tube used by Reynolds.

L. J. Briggs (1950) carried out similar experiments using a centrifugal method, and he also studied the effect of temperature on the breaking tension in the range 0° to 50°C. In his work the liquid was held in a Z-shaped capillary tube open at both ends; the Z-tube was mounted horizontally and spun in its plane about a vertical axis through its centre. The tube was enclosed in a cylinder and the air pressure in the cylinder when operating was reduced to about 4 mm Hg. As in Reynolds' work, the breaking tension could be calculated from the value of ω which caused the liquid column to break at its centre. The results Briggs obtained for distilled water in a Pyrex glass capillary show that the tension had a maximum value of 277 atm around 10°C and that it decreased to 217 atm as the temperature rose to 50°C. This decrease is to be expected because at the critical temperature the tension must be zero. As the temperature fell from 5° to 0°C the breaking tension fell very rapidly, and this represents another anomaly in the behaviour of water in this temperature region.

Briggs emphasizes that with this method it cannot be said whether the rupture of the column occurs on the wall of the capillary (loss of adhesion) or in the body of the liquid (loss of cohesion).

8.3 DYNAMIC STRESSING OF A LIQUID

In the static experiments we have considered in the previous section, a gradually increasing tension was applied over a comparatively long interval of time, of the order of half an hour; that is, we had a fairly low rate of stressing. We now turn to the case where the tension is applied over a much shorter time

interval, less than a millisecond perhaps. Much of this work has been in connection with the phenomenon known as **cavitation**. This is the name given to
the formation of small cavities in the liquid occurring when the tension falls to
the breaking value. These cavities are thought to contain vapour of the liquid
and dissolved gases. Cavitation does occur in the static Berthelot tube experiments, but its detailed study has been more in connection with work on
dynamically applied tensions and ultrasonic stressing. We shall consider this
phenomenon in greater detail in the next section.

Couzens and Trevena (1969, 1974) carried out experiments in which a
pulse of compression was propagated upwards through a liquid column. After
reflection at the free surface of the liquid, it travelled downwards as a pulse of
tension. The details of the apparatus are briefly as follows. The liquid was contained in a vertical cylindrical stainless steel tube fitted with a piston at its lower
end and with its upper end open. A pressure pulse was produced in the liquid by
firing a lead bullet so as to strike the lower end of the piston normally and
centrally. The variation of pressure with time at a point in the liquid at a depth
x below the free surface was followed by a piezoelectric pressure transducer
mounted in the wall of the tube.

In the upward-going pressure pulse, the pressure rose to a maximum value
p in about 50 μs and then decayed more gradually to zero, the whole pulse
having a total duration τ of about 500 μs. If c denotes the velocity of propagation of this pulse in the liquid, then provided $2x/c > \tau$ the reflected tension
pulse would not overlap the incident pulse in time. This condition was satisfied
by ensuring that the depth x was sufficiently large. Let F denote the maximum
tension in the reflected pulse. The procedure then consisted of increasing p

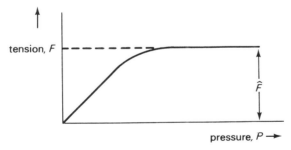

Fig. 8.2   Plot of F against p, showing the limiting value of $\hat{F}$. [After Trevena].

gradually by using different bullets and different pistons of various lengths.
In each case both the maximum pressure p in the upward pulse and the maximum tension F in the down-going reflected pulse were recorded. A plot of F
against p was then made. In all cases it was found that the value of F tended to
an upper limiting value $\hat{F}$ as the value of p was increased (see Fig. 8.2). This
shows quite clearly that $\hat{F}$ is the maximum possible tension that can be sustained

by the liquid under these conditions.

For ordinary tap water this breaking tension was $\hat{F} = 8.5$ atm, and for boiled de-ionized water $\hat{F} = 15.0$ atm. The higher value in the second case is to be expected, because the boiling would remove much of the dissolved gas and would make it easier for the liquid to withstand a higher tension.

In conclusion it should be noted that the critical tension values obtained under these rapid rates of stressing appear to be consistently lower than those obtained statically.

8.4　ULTRASONIC STRESSING OF A LIQUID

This type of stressing accompanies the propagation of ultrasonic waves in a liquid. A liquid is subjected to a compression and a tension during the respective positive and negative half-periods of the pressure cycle when such a high-frequency sound wave is passed through it. As the tension increases from zero in the negative half-period a point will be reached when cavitation sets it. As the pressure subsequently increases in the positive sense, the cavitation bubbles contract rapidly and often disappear completely.

It is probably true to say that of all the effects associated with the propagation of ultrasonics in a liquid, the phenomenon of cavitation is the best known but the least understood. In particular the mechanism of the initiation of cavitation is still not clear. It can set in either at the wall of the vessel containing the liquid or in the body of the liquid itself, but the factors involved are probably quite different in the two cases.

There are in general two main types of cavitation bubble. The first is the gas-filled bubble containing gas (air) previously dissolved in the liquid; such bubbles grow to visible size and reach a state of considerable stability. The stage at which these bubbles appear is often referred to as **degassing**. The second type of bubble is vapour-filled, containing the vapour of the surrounding liquid; such bubbles are much smaller than the gas-filled type and their production and collapse are of a much more explosive nature. A third type of bubble in which the bubble is practically a complete void is also postulated by some workers.

Blake (1949a,b) attempted to describe the behaviour of gas-filled cavities by invoking a process of **rectified diffusion**. During the positive half-cycle gas diffuses out of the bubble and during the negative half-cycle gas flows back in. However, because the surface area of the bubble is larger during the negative half-cycle and the oscillation is asymmetrical over a large number of cycles there is a net inflow of gas. This process explains the growth of gas-filled bubbles in the ultrasonic field. Willard (1953) extended some of the theoretical work of Blake and obtained photographs of the growth and decay of cavitation bubbles in degassed and aerated water. He estimated that most of the cavitation occurred at peak tensions of 20 atm or less.

H. B. Briggs and his colleagues (1947) used high-amplitude sound waves of frequency 25 kHz in their work. With liquids containing dissolved air they found that cavitation set in when the peak negative acoustic pressure reached a value equal to the ambient pressure, that is to atmospheric pressure. On the other hand, they found that with degassed liquids the peak negative pressure needed to produce cavitation was much greater.

8.5 THEORETICAL WORK ON THE TENSILE STRENGTH OF LIQUIDS

Temperley (1947) has shown how the van der Waals' equation may be used to estimate the tensile strength of a liquid.

Let us assume that we have a van der Waals' liquid and consider two iso-thermals corresponding to temperatures T_1 and T_2 as in Fig. 8.3. The positions of the horizontal dotted lines AD and EH are determined by the rule of equal areas (see Appendix 1), and the portions AB, EF correspond to metastable regions (see Chapter 2.2). For both curves the rule of equal areas gives a positive vapour pressure, but at the lower temperature T_2 the part E'F of the metastable region corresponds to the liquid being at a negative pressure, that is, under tension. It is then argued that the point F, which represents the limit of the metastable region at this temperature, must also represent the limiting value of the tension that can be sustained by the liquid. In other words the ordinate OF' represents the **tensile strength** of the liquid. Extending this argument we could say that whatever equation of state we consider in general the limit of the tensile strength corresponds to a condition such as that existing at the minimum at F, that is, $dP/dV = 0$. Temperley then argues that some estimate of the tensile strength can be obtained using this theory, even though water does not obey the van der Waals' equation very well. Using suitable values for the van der

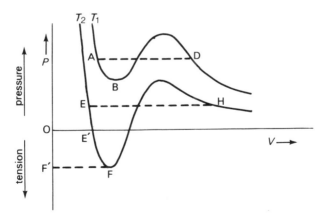

Fig. 8.3 Van der Waals' curves illustrating Temperley's theory.

Waals' constants a and b and putting T_2 equal to room temperature, the value of the tensile strength of water turns out to be of the order of 500 atm. As we have seen, experimentally it is not the true tensile strength which is measured but rather the value of the breaking tension corresponding to the weakest link in the system, the value reached just before the liquid tears away from the walls of the containing vessel. Although this theoretical value of the tensile strength given by Temperley is much greater than the experimental values of the breaking tension (with the exception of L. J. Briggs' centrifugal values for water) it is nevertheless in far better agreement with experiment than theoretical values which have been given by other workers. For example, Fürth (1941) gave a theoretical value of 3000 atm for the tensile strength of water. Other workers have suggested that the numerical value of the tensile strength is the same as the **intrinsic pressure** a/V^2 in the van der Waals' equation, which is of the order of 10,000 atm.

A theoretical investigation of the amount by which the tensile strength of a liquid should be reduced by a dissolved gas has been made by Kuper and Trevena (1952). They considered a gas whose van der Waals' constants are a_1, b_1, being dissolved in a liquid with corresponding constants a_2, b_2 and they treated the whole system as a binary mixture. Their results showed that the reduction to be expected in the tensile strength of water caused by saturating it with air at a pressure of 1 atm is less than 0.5 per cent. At first sight the smallness of this decrease is surprising, yet on closer investigation it does not seem to be in disagreement with the experimental evidence. For example, Ursprung (1915) and Renner (1915) gave values of 300 atm for the tensile strength of cell sap which was probably nearly saturated with dissolved air. On the other hand, as we have seen, there does seem to be definite evidence that the breaking tension as measured by ultrasonic methods increases as the gas is removed. This increase is probably due in part to the removal of undissolved gas nuclei and in part to the process of rectified diffusion mentioned earlier. Thus a progressive diffusion of gas into microscopic bubbles occurs in forced oscillation in the ultrasonic field. Such a mechanism is absent when the stress is purely hydrostatic.

8.6   OTHER EXPERIMENTAL WORK

Hayward (1964) has been able to obtain a part of the (P, V) curve for a liquid in the negative pressure region. The liquid under test (mineral oil) was contained in a Berthelot tube of hardened steel with a screw cap at either end and tension was generated by first heating the tube and then cooling it as before. This tension was measured by mounting strain gauges on the tube and also on an identical tube with caps removed, both tubes being in the same constant-temperature bath. These two strain gauges formed two arms of a bridge circuit and the unbalance in the bridge was a measure of the pressure (positive or negative) in the sealed tube. Volume changes, as in previous Berthelot tube work, were calculated from the thermal and elastic constants of the steel and liquid. A portion of the (P,V) curve in the negative pressure region was obtained. This portion was curved and the curvature increased with increasing negative pressure, that is, tension. This strongly suggests that, for even greater values of the tension, the curve would exhibit a minimum given by $dP/dV = 0$. This point would correspond to the true tensile strength of the liquid.

Recently Richards and Trevena (1976) have refined the Berthelot tube method by using a strain-gauge pressure transducer built into one end of the tube. This follows directly the pressure and tension variations with temperature in the enclosed liquid. A family of curves which are virtually true isochores (that is, 'constant volume' curves) was obtained for deionized water, and it is believed that this is the first time that such curves have been reported in the *negative* pressure region (see Fig. 8.4). These curves extend to negative pressures of about -30 atm but, because of the fracture of the liquid inside the Berthelot tube, it was not possible to follow the curves for any lower values of these pressures.

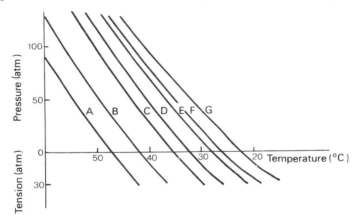

Fig. 8.4 A family of experimental (pressure, temperature) curves obtained for deionized water in a steel Berthelot tube. [After Richards and Trevena].

Apfel (1971) has used a rather novel way to measure the tensile strength of liquids. A droplet of one liquid A was injected into a much larger volume of a second liquid B; the two liquids did not mix or otherwise interact. The great advantage of this arrangement was that the 'container' of the test liquid A was the perfectly smooth surface of the liquid B. Consequently there could be no possibility of cavitation being initiated at the solid walls of any containing vessel — a process which so often bedevils tensile strength measurements. The droplet was acoustically stressed in a standing-wave field, and at a sufficiently high amplitude the droplet reached the limit of its tensile strength and an explosive liquid-to-vapour transition occurred.

An elegant method, in which an explosive charge was detonated a small distance below the surface of a liquid, has been used by Wilson, Hoyt and McKune (1975). The basic idea was to obtain high-speed motion photographs of the spray dome formed above the original undisturbed surface as a result of detonating this charge. From these photographic records they obtained a value of 8.0 atm for the breaking tension of ordinary water, in close agreement with the value of 8.5 atm obtained by Couzens and Trevena.

Recently Sedgewick and Trevena (1976) have used the method of dynamic stressing described in Chapter 8.3 to investigate how the breaking tension of ordinary tap water varies between 0 and 70°C. Their results are shown in Fig. 8.5, where the breaking tension is plotted against the temperature of the water. This graph shows that the breaking tension is at a maximum when the temperature is around that corresponding to the maximum density of water. At this maximum density the molecules are in their most closely packed configuration, and one would intuitively expect the tension needed to 'pull them apart' to be greatest in this situation.

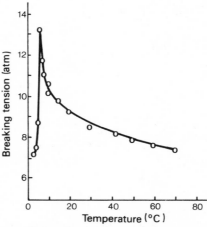

Fig. 8.5 Graph showing the variation of breaking tension with temperature [after Sedgewick and Trevena].

A recent way of studying the dynamics of a single cavitation bubble is to irradiate a liquid by a focused high-power laser beam. Accounts of such methods have been given by Lauterborn (1973) and by Felix and Ellis (1971).

Carlson and Henry (1973) have subjected a liquid to tension at extremely high stress rates. The essential part of their apparatus is shown schematically in Fig. 8.6. The liquid L was confined between a solid plate S and a stretched aluminized Mylar film M. A pulsed electron beam generator was used to produce a stress pulse in the solid plate. This compressional stress wave generated in the solid travelled into the liquid and, because the acoustic impedances of the solid and liquid were chosen to be virtually the same, it was only very slightly affected in crossing the solid-liquid boundary. After travelling through the liquid, the compressional wave was reflected as a wave of tension at the interface M. In the work reported, the solid plate used was made of a composite organic material known as Astrel 360 and its thickness was 2 mm. The liquid used was glycerol and the sample had a thickness of 3 mm, while the stretched Mylar film was only 6 μm in thickness.

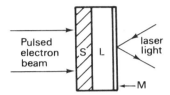

Fig. 8.6 Apparatus used by Carlson and Henry.

The free-surface motion of the Mylar film caused by the reflection of the stress wave incident on it from the liquid was monitored by means of a velocity interferometer, using light from a He-Ne laser. From an observation of fringe shift data the (velocity, time) records of the Mylar film could be obtained, and hence a value of the negative stress that was generated in the liquid. From a total of 14 experiments Carlson and Henry concluded that the average tensile strength of glycerol was 0.6 kbar, about 600 atm. They also make the point that the glycerol did break in each of the 14 attempts, and this failure occurred at a distance of about 0.2 mm away from the Mylar film. For further details the reader is referred to two papers by Trevena (1967, 1976).

8.7 INTRODUCTION TO LIQUIDS UNDER COMPRESSION

We have already seen in Chapter 2 that a liquid cannot support a static shear stress. It therefore follows that the only modulus that can be defined for a liquid is its bulk modulus, K. The compressibility of a liquid is usually small and numerically comparable with that of a solid. Note that the consequences of treating a liquid as incompressible were examined in Chapter 6. If we consider a volume V of liquid and then subject it to a hydrostatic pressure ΔP, the volume will change to $V - \Delta V$ and the volume strain will be $-\Delta V/V$. The bulk modulus is then

$$
K = \frac{\Delta P}{-\Delta V/V}
$$

$$
= -V \frac{\Delta P}{\Delta V} .
$$

The value of the bulk modulus will depend on the rate at which the pressure changes are carried out. If the pressure is applied slowly, the liquid will remain at a constant temperature provided it is in a vessel with conducting walls; under these conditions we have the *isothermal* bulk modulus, K_T. If, however, the pressure changes are so rapid that virtually no heat exchange with the containing vessel is possible in the time available, the *adiabatic* bulk modulus K_S is obtained.

The isothermal compressibility κ_T is defined as the reciprocal of the iso-thermal bulk modulus and is written as

$$\kappa_T = -\frac{1}{V}\left(\frac{\partial V}{\partial P}\right)_T \, ,$$

where the suffix T emphasizes the fact that we are considering isothermal conditions. It is this isothermal compressibility (or its reciprocal) which is usually involved in the measurement of PVT data for liquids.

Canton (1762) first demonstrated that liquids were compressible by observing the movement of a liquid in a capillary attached to a bulb filled with the liquid. During the two centuries since Canton's early work the physics of high pressures has been slowly developed. Even today really accurate measurements are difficult, and the reader is referred to a report by Isdale, Brunton and Spence (1975) for a critical survey of available methods. The most outstanding name in the field is that of Bridgman (1949) in whose book and papers a comprehensive account of this work is available. We shall give a brief summary of this work in the next section.

8.8   THE EXPERIMENTS OF BRIDGMAN AND OTHERS

In measuring the bulk modulus of a liquid, there are two particular difficulties involved. First, the change of volume of the liquid when compressed is very small. Secondly, the expansion of the containing vessel is comparable with the change of volume of the liquid. Even if the same pressure were applied inside and outside the vessel, the internal volume of the vessel would not remain constant. It would decrease by the same amount as if a piece of the solid that exactly fitted the inside of the vessel were subjected to the same pressure on its outside. The expected volume change of a vessel can be estimated theoretically for vessels of simple geometrical shapes such as spheres or cylinders with flat ends. Metal vessels are best used because they are elastically isotropic.

In Bridgman's work the test liquid was placed in a strong alloy-steel cylinder and the apparent change in volume of the liquid relative to the cylinder measured by the advance of the piston producing the pressure. Bridgman ensured that the liquid did not leak upwards by using a special packing arrangement (see Fig. 8.7) As it moves down, the piston A transmits the pressure via the steel ring B to the piston C through soft rubber packing kept in place by copper washers D. Since the total upward force exerted by the test liquid on the lower face of C must equal that exerted by the packing the pressure in the packing is automatically greater than that in the liquid which therefore cannot leak. The pressure was determined by measuring the change in resistance of manganin wire contained in the cylinder. A correction for the expansion of the cylinder itself could be made by further experiments in which part of the liquid was replaced by a steel plug. With this apparatus Bridgman obtained (P, V) isotherms over a wide temperature range.

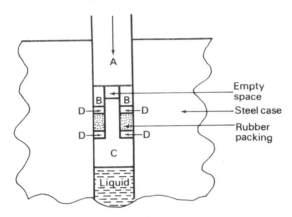

Fig. 8.7 Bridgman's apparatus.

Bridgman also used a method in which the liquid was contained in a flexible bellows. As the liquid was compressed, the bellows contracted and moved a slide-wire AB which carried a constant current. An electrical contact like a potentiometer jockey key touched the wire at some point C along its length and a relation between the pressure and the potential differences across the portions AC and AB was obtained. The ratio of these p.d.'s equals the ratio of the lengths, that is AC/AB, from which the volume changes could be calculated.

A less direct method of measuring the bulk modulus of a liquid is from measurements of ultrasonic velocity, c, using the equation $K_S = \rho_0 c^2$. This gives the adiabatic bulk modulus K_S, from which the isothermal bulk modulus K_T can be obtained by using the thermodynamic relationship

$$\kappa_T = \kappa_S + \frac{Tv\,\beta^2}{C_p}$$

between the two compressibilities. In this equation v is the specific volume, β the volume coefficient of expansion and C_p the heat capacity per unit mass.

Isdale *et al* (1975) follow up and extend the critical analysis by Hayward (1971) of the various methods which have been used to measure the compressibilities of liquids. Hayward found discrepancies of 10% (and sometimes much more) between the values of the same compressibility measured by different workers under the same conditions of temperature and pressure. He concluded that these discrepancies are mainly due to anisotropic distortion of the apparatus and to the presence of entrained air in the system. Previous to Hayward's work the ultrasonic method probably yielded the best values of the isothermal compressibility.

8.9   DEDUCTIONS FROM HIGH PRESSURE WORK ON LIQUIDS

The most obvious effect produced by the application of pressure to a liquid is the volume change. The way in which volume changes with pressure is of fundamental importance because it leads to PVT data, and so to the equation of state. The various experimental results yield a series of (P, V) relations at constant T, the isotherms. Various empirical equations relating P and V at constant T have been compared with these data; two examples are the Tait equation (1888) and the Hudleston equation (1937). If a sufficient number of isotherms is available then isobars $[(V, T)$ curves at constant $P]$ and isochores $[(P, T)$ curves at constant $V]$ can also be plotted.

If the various PVT data are combined with the results from ultrasonic velocity measurements, some information about the intermolecular forces can be obtained from thermodynamics. Consider the equation

$$\left(\frac{\partial U}{\partial V}\right)_T = T\left(\frac{\partial P}{\partial T}\right)_V - P.$$

Now both T and $(\partial P/\partial T)_V$ are positive and so for small values of P the right-hand side is positive; this means that $(\partial U/\partial V)_T$ is positive. Thus the internal energy U decreases as the volume decreases due to the applied pressure P. As P increases the right-hand side of the equation eventually becomes zero, after which it becomes negative. At the point where $(\partial U/\partial V)_T = 0$, the intermolecular forces change from being attractions to being repulsions in keeping with the known form of interactions between molecules.

The compressibility of all liquids decreases with increase of pressure, at first rapidly and then more slowly as the pressure becomes higher. The initial high value of the compressibility occurs because of a decrease in the intermolecular spacing. The smaller values at the higher pressures are thought to be due to a decrease in volume of the actual molecules themselves once they are nearly in contact with their neighbours. The compressibility of most liquids increases as the temperature rises, and again this is what we might have expected. As the temperature rises collisions between molecules become more vigorous and, because the molecules are not completely rigid, their effective volume will decrease.

The theoretical value of the bulk modulus turns out to be very sensitive to small changes in the molecular distribution function of a liquid. This is so particularly at small distances from the central molecule — the very region that is the hardest to measure experimentally or predict theoretically! Therefore bulk modulus measurements are not very helpful in checking liquid state theory, though they are of very considerable practical importance. Nearly every industrial process involves the liquid state, very often at high pressures. Obvious examples are processes involving servo-mechanisms (e.g. the hydraulic press), chemical preparations at high pressures and the study of the flow of oil in porous rocks. All of these need accurate knowledge of the equation of state. Most of the experiments can be represented by a virial series (similar to that of an imperfect gas) truncated at about the third term.

For a discussion of the effects of high pressure on the transport, chemical and biological properties of liquids, the reader is referred to *High Pressure Physics and Chemistry*, (1963).

REFERENCES

Books

Bradley, R. S. (editor), 1963, *High Pressure Physics and Chemistry,* Vol. 1, Academic Press, London and New York.

Hamann, S. D., 1957, *Physico-Chemical Effects of Pressure,* Butterworths Scientific Publications, London.

Bridgman, P. W., 1949, *The Physics of High Pressure,* G. Bell and Sons, London.

Papers

Apfel, R. E., 1971, *J. Acoust. Soc. Amer.,* **49**, 145.

Berthelot, M., 1850, *Ann. Chim. Phys.,* **30**, 232.

Blake, F. G., 1949a, *J. Acoust. Soc. Amer.,* **21**, 464.

Blake, F. G., 1949b, *Tech. Memor. Harvard Univ. Acoust. Res. Lab.,* No. 12.

Briggs, H. B., Johnson, J. B. and Mason, W. P., 1947, *J. Acoust. Soc. Amer.* **19**, 664.

Briggs, L. J., 1950, *J. Appl. Phys.,* **21**, 721.

Couzens, D. C. F. and Trevena, D. H., 1969, *Nature,* **222**, 473.

Couzens, D. C. F. and Trevena, D. H., 1975, *J. Phys. D.,* **7**, 2277.

Felix, M. P. and Ellis, A. T., 1971, *App. Phys. Letters,* **19**, 484.

Fürth, R., 1941, *Proc. Camb. Phil. Soc.,* **37**, 276.

Hayward, A. T. J., 1964, *Nature,* **202**, 482.

Hayward, A. T. J., 1971, *J. Phys. D.,* **4**, 938.

Hayward, A. T. J., 1971, *J. Phys. D.,* **4**, 951.

Isdale, J. D., Brunton, W. C and Spence, C. M., 1975, *National Engineering Laboratory Report,* No. 591 (this report includes a bibliography).

Kuper, C. G. and Trevena, D. H., 1952, *Proc. Phys. Soc. A,* **65**, 46.

Rees, E. P. and Trevena, D. H., 1964a, *Brit. J. Appl. Phys.*, **15**, 337.

Rees, E. P. and Trevena, D. H., 1964b, *Nature*, **203**, 396.

Reynolds, O., 1878, *Mem. Manch. Lit. Phil. Soc.*, **17**, 159.

Richards, B. E. and Trevena, D. H., 1976, *J. Phys. D.*, **9**, L123.

Sedgewick, S. A. and Trevena, D. H., 1976, *J. Phys. D.*, **9**, 1983.

Temperley, H. N. V. and Chambers, Ll. G., 1946, *Proc. Phys. Soc.*, **58**, 420.

Temperley, H. N. V., 1946, *Proc. Phys. Soc.*, **58**, 436.

Temperley, H. N. V., 1947, *Proc. Phys. Soc.*, **59**, 199.

Trevena, D. H., 1967, *Contemp. Phys.*, **8**, 185.

Trevena, D. H., 1976, *Contemp. Phys.*, **17**, 109.

Ursprung, G., 1915, *Ber. deutsch. Bot. Ges.*, **33**, 153.

Willard. G. W., 1953, *J. Acoust. Soc. Amer.*, **25**, 669.

Wilson, D. A., Hoyt, J. W. and McKune, J. W., 1975, *Nature*, **253**, 723.

Chapter 9
SURFACE PHENOMENA

9.1 INTRODUCTION

We are all familiar with the fact of capillary rise, the rise of a liquid, such as water, in a capillary tube whose lower end is dipped into the free surface of the liquid. We also know that mercury can form small droplets whose shape becomes more spherical as their size decreases, and that a drop of one liquid placed on the surface of a second liquid will spread over this second liquid surface. These phenomena show clearly that there are forces other than gravity acting on the liquid, and they are best described as capillary phenomena or capillarity, though they have usually been studied under the name of 'surface tension'. This term 'surface tension' is less used nowadays and is being replaced by the term **free (or mechanical) surface energy**. The surface energy concept makes it easier to interpret capillary phenomena in terms of the inter-molecular forces in a liquid.

We have already seen that the force-separation curve between two molecules at distance r apart is as shown in Fig. 9.1. At small values of r there is a strong repulsion; as r increases beyond the equilibrium separation r_0 the force becomes

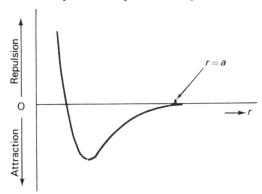

Fig. 9.1 (Force, separation) curve for two molecules.

attractive, quickly reaches a maximum and almost as quickly falls effectively to zero at about distance a. This distance a around a molecule is only a few molecular diameters (see Chapters 3.2 and 3.5).

Over a given period, the time spent in collisions between molecules (that is, when repulsive forces are operative) is much less than that during which the molecules are farther apart (when attractive forces are operative). Consider then a molecule A inside the bulk liquid (Fig. 9.2). Let us draw a sphere of radius a with A as centre. According to the above discussion, the molecules outside this

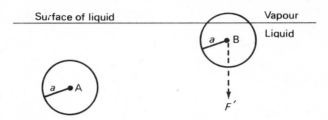

Fig. 9.2 Illustrating the 'sphere of influence'.

sphere will not exert any attraction on A. However molecules within the sphere − of the order of a few hundred in number at most − will exert forces on A, those due to the nearest neighbours repulsive, but mostly attractive due to the farther neighbours. The sphere may therefore be termed the **sphere of influence** of the molecule A.

The actual resultant force on A will vary with time because of collisions and the movement of the molecules near A, but the time-average of this force (over a time much greater than the average time between collisions) will be zero. Thus for any molecule, like A, situated at a distance greater than a from the surface, the time-average of the force on it is zero. This distance a is of the order of 10^{-9} m, and any molecule at a greater distance from the surface may be said to be in the 'bulk' liquid. For example, Tabor reports that there is recent experimental evidence to show that a water film of thickness 5×10^{-9} m exhibits 'bulk' properties.

Now consider a molecule B whose distance from the surface is less than a (Fig. 9.2). A part of its sphere of influence will then be above the surface where there is a relatively small concentration of vapour molecules (except, of course, near the critical temperature). The molecules which are *nearest* neighbours of B will, depending on their distance from B, exert either repulsions or attractions on B. However the resultant force on B due to these nearest neighbours will be zero as a time-average. On the other hand, when we consider the molecules which are farther away from B (and these all exert *attractions* on B) we see that there are more of these below B than above it. Thus the time-average F' of the steady force on B will be downwards and normal to the surface. The nearer B is to the surface the greater the magnitude of F'. In fact F' will be a maximum at the surface and become effectively zero at distance a below it. Molecules in this 'anomalous' surface layer of thickness a are therefore subject to an inward force and so tend to be accelerated inwards from the surface. This explains why the number of molecules in the surface, and hence the surface area itself, tends to decrease. Of course this process of contraction is eventually opposed by the repulsive forces between the molecules which prevent a complete inward 'collapse' of the liquid. Once the state of minimum surface area is achieved, an equilibrium is set up in which the molecules leaving the surface inwardly are continually replaced by others 'dredged up' from beneath.

Let us now, following Feather (1959), consider a molecule inside the bulk liquid and refer to Fig. 9.3. If this molecule is moved towards the surface, no work is done on it until it reaches a distance $OL = a$ from the surface. As it

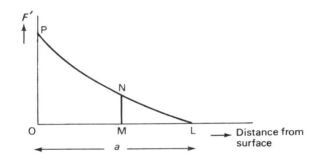

Fig. 9.3 How the surface of a liquid is the seat of potential energy.

approaches nearer the surface it is opposed by a gradually increasing force F', and hence work must be done on it to bring it nearer the surface. For example, the work done in moving it from the top of the bulk liquid (depth OL) to a depth OM in the surface layer is $\int_{r=OL}^{r=OM} F' dr$ and is represented by the area MLNM. This also represents the increase in the potential energy of the molecule in raising it. Similarly the increase in potential energy on moving the molecule from the bulk liquid to the surface is represented by the area OLPO. Thus the surface of the liquid can be regarded as the seat of potential energy. Increasing the surface area means bringing more molecules to the surface with a corresponding increase in the potential energy.

There is an alternative way of looking at the energy change involved in bringing a molecule from the bulk liquid to the surface, by considering the bonds between the molecule and its nearest neighbours. To break such a bond requires energy so that the energy of a molecule increases with a decrease in the number of bonds it forms. Now a molecule in the surface has about half as many such bonds as a molecule in the bulk liquid, thus the surface molecule has the higher energy (see Chapter 9.3).

9.2 SURFACE ENERGY AND SURFACE TENSION

To illustrate these concepts let us consider two simple facts. First suppose that a loop of cotton is embedded in a soap film in a rectangular frame as in Fig. 9.4a. When the film inside the loop is broken it is found that the thread is pulled out into a perfect circle as in Fig 9.4b. Since a circle is the curve enclosing

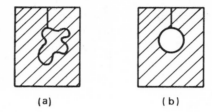

(a)   (b)

Fig. 9.4   A soap film contracts to a minimum area.

the largest area for a given perimeter, this shows us that the soap film outside the cotton is tending to contract to as small an area as possible. This experiment suggests that the contraction of a surface is associated with a decrease of potential energy and vice versa. Thus we are led to define the **free (or mechanical) surface energy**, γ, as the work that must be done to increase the surface area by unity under isothermal conditions. Hence γ has the units $\mathrm{Jm^{-2}}$.

Secondly consider the formation of a drop of pure water on a cylindrical tube. The shape of the drop just before it breaks away is as shown in Fig. 9.5. If we consider the horizontal dotted plane, the weight of the drop below this

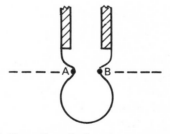

Fig. 9.5 Formation of a drop on a cylindrical tube.

may be regarded as being balanced by vertical forces acting tangentially to the surface of the liquid around the horizontal circle of diameter AB. In this way we are led to the idea of two regions of a liquid exerting surface forces on each other in a direction perpendicular to the line separating them. The force per unit length acting perpendicular to such a line is known as the **surface tension**, which is measured in $\mathrm{Nm^{-1}}$. Note that both the free surface energy and the surface tension have the same dimensions of mass divided by $(\text{time})^2$.

In deriving various important results in capillarity either the surface tension or the surface energy approach can be used. They are both formally equivalent, but the energy approach is preferable because it is more fundamental in its physical explanation of capillary phenomena (see Yarnold, 1960). Indeed, in his classic book, *The Physics and Chemistry of Surfaces,* N. K. Adam writes:

> 'This free energy in the surface is of fundamental importance; a vast number of problems relating to the equilibrium of surfaces can be solved without knowing more than the magnitude of this free energy. In the solution of such problems a mathematical device is almost invariably employed to simplify the calculations; it is to substitute for the surface free energy a hypothetical tension, acting in all directions parallel to the surface, equal to the free surface energy. This is what is generally known as the *surface tension*. It is always possible, mathematically, to replace a free energy per unit area of surface by a tension acting parallel to the surface.'

This equivalence between the free surface energy and a tension acting in the surface may be further illustrated by a film on a wire frame, one of whose arms AB can move without friction (Fig. 9.6). Thus a force as shown must be applied to AB if the contraction of film surface to a state of lower mechanical energy is to be prevented. Let F (which is normal to AB and parallel to the surface) be the force which is just adequate for this purpose. The arm AB is now moved slowly, and hence isothermally, through a distance x. The surface area increases by $2\ell x$ because the film has two surfaces,

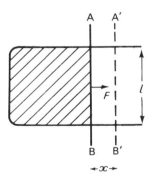

Fig. 9.6 The numerical equivalence of surface tension and free surface energy.

and so the work done in increasing this area is $2\gamma\ell x$. This must be equal to $-Fx$ the work done against the external force. Thus, by the principle of virtual work,

$$2\gamma\ell x - Fx = 0$$

or
$$\gamma = F/2\ell.$$

Because of the two sides of the film the mechanical force F acts on a total line distance of 2ℓ in the film. Thus γ is also the mechanical force per unit length, that is the surface tension, which is thus numerically equal to the free surface energy.

When any change of surface area occurs, the *total* change in energy of the surface is not usually equal to the change in the free surface energy. The explanation for this lies in the fact that γ decreases as the temperature rises. If the surface is increased isothermally the liquid will absorb heat from its surroundings, if it is enlarged adiabatically it will cool. When a surface changes in area from A to $A + dA$ in a reversible way, the following equation holds:

$$TdS = C_A dT - T\left(\frac{\partial \gamma}{\partial T}\right)_A dA . \tag{9.1}$$

This result is proved, for example, in Zemansky's *Heat and Thermodynamics*. Here C_A is the heat capacity of the liquid at constant surface area A.

As a result of the area change the work done on the surface is $dW = \gamma dA$. If we apply the First Law of Thermodynamics to this change we have

$$dQ = dU - dW = dU - \gamma dA . \tag{9.2}$$

where dQ is the heat transferred to the liquid form its surroundings and dU the change in the internal energy of the liquid. If the process is isothermal then $dT = 0$, and if we assume that γ is a function of temperature only and does not depend on A then (9.1) becomes

$$TdS = -T \frac{d\gamma}{dT} dA . \tag{9.3}$$

But $dQ = TdS$ and so from (9.2) and (9.3)

$$-T \frac{d\gamma}{dT} dA = dU - \gamma dA$$

or
$$dU = \left(\gamma - T \frac{d\gamma}{dT}\right) dA .$$

If we consider a finite change in the area from an initial value A_1 to a final value A_2, with U correspondingly changing from U_1 to U_2, then

$$\frac{U_2 - U_1}{A_2 - A_1} = \gamma - T \frac{d\gamma}{dT} .$$

If A_1 is very small and $A_2 \gg A_1$, the left hand side of the above equation is interpreted as E, the total surface energy per unit area. Thus

$$E = \gamma - T \frac{d\gamma}{dT} . \qquad (9.4)$$

For water at room temperature, γ is about 72×10^{-3} Jm^{-2} ($= 72 \times 10^{-3}$ Nm^{-1}). It must be emphasized that this is the value of γ when the water is in contact with its vapour, that is it is the value for a liquid-vapour interface and the effect of the atmosphere is negligible. The value of γ for a water-chloroform interface at room temperature for example is about 28×10^{-3} Jm^{-2}. All this serves to underline the fact that when the second medium is not the liquid vapour but another liquid — or a solid — this fact must be clearly stated.

Returning to equation (9.4) we see that when a surface area increases isothermally, the total surface energy E is made up of two parts, (a) the free surface energy change γ due to the enlarging of the surface and (b) 'the latent heat' term $-T \frac{d\gamma}{dT}$ which must be provided to keep the temperature constant. Since $d\gamma/dT$ is negative this 'latent heat' term is actually positive. For water in contact with its vapour $\gamma = 72 \times 10^{-3}$ Jm^{-2} and E is 118×10^{-3} Jm^{-2}.

9.3   SURFACE ENERGY AND NEAREST–NEIGHBOUR BONDS

On average, a molecule in the bulk liquid is surrounded by z nearest neighbours each at a distance of about r_0 from it. In the surface each molecule has $z/2$ nearest neighbours. Consequently to create a new liquid surface some nearest-neighbour bonds will have to be cut. Suppose therefore that a liquid column of unit cross-section is broken by some means so as to produce two new surfaces S_1 and S_2 each of unit area (Fig. 9.7). Let there be n molecules in the cross-section of the liquid column. After the break each molecule in both

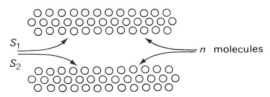

Fig. 9.7 Creating two new surfaces S_1 and S_2 by breaking nearest-neighbour bonds.

surfaces S_1 and S_2 will have, on average, $z/2$ nearest neighbours and not z as before the break. Hence the creation of the two new unit surfaces means that $\frac{1}{2}nz$ nearest-neighbour interactions have been cut, requiring an amount of energy $\frac{1}{2}nz\epsilon$ (see Fig. 3.1b). So the energy needed to produce a *single* new unit area of surface, that is the surface energy, is $nz\epsilon/4$. Since n is also about $1/\sigma^2$, where σ is the diameter of a molecule (see for example *Properties of Matter,* Chapter 3, by Flowers and Mendoza, or *Gases, Liquids and Solids,* Chapter 7, by Tabor) then $z\epsilon/4\sigma^2$ is a close estimate for the value of the surface energy. Of course, it is assumed that the two unit surfaces in Fig. 9.7 are created isothermally and reversibly.

9.4   INTERFACES AND YOUNG'S RELATION

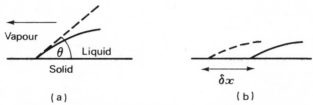

Fig. 9.8 Diagrams used in derivation of Young's relation.

Consider a liquid in contact with a plane solid surface as in Fig. 9.8a. There are three phases involved, solid, liquid and vapour, and hence three interfaces, solid-liquid, liquid-vapour and solid-vapour. To these interfaces there will be free surface energies of γ_{SL}, γ_{LV} and γ_{SV} respectively. In the diagram an angle θ, called the angle of contact, is shown. This is the angle between the tangential planes to the solid-liquid and liquid-vapour interfaces as shown. We now derive a relation between the three free surface energies and θ.

For this purpose we suppose that the liquid in Fig. 9.8a extends to infinity over the plane surface of the solid to the right of the line of contact. Consider now a movement of the liquid to the left in the direction of the arrow in Fig. 9.8a so that the line of contact moves through a small distance δx as in Fig. 9.8b. In this case the overall potential energy will not change. Three area changes will be involved if we consider unit length of the line of contact perpendicular to the plane of the diagram of Fig. 9.9 namely:

Area increase of solid-liquid interface $= \delta x$

Area decrease of solid-vapour interface $= \delta x$

Area increase at liquid-vapour in interface $= \delta x \cos\theta$.

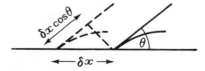

Fig. 9.9 Area changes involved in deriving Young's relation.

By the principle of virtual work the total change in potential energy is zero and so

$$\gamma_{SL}\delta x + \gamma_{LV}\delta x \cos\theta - \gamma_{SV}\delta x = 0$$

or
$$\gamma_{SV} = \gamma_{SL} + \gamma_{LV}\cos\theta . \qquad (9.5)$$

This is the general relation between the three interfacial free surface energies. It is known as Young's relation after T. Young who gave it in 1805. It was derived by A. Dupré in 1869.

In a situation like that shown in Fig. 9.8a the angle θ will depend on the nature of the solid and the liquid, and to a much less extent on the vapour. We can distinguish between three cases:

(i) $0° < \theta < 90°$. An example is water in contact with a slightly greasy glass surface. This is referred to as 'imperfect wetting' or 'partial wetting'.

(ii) $\theta = 0°$. In certain cases where there is a strong attraction between the liquid and solid molecules the angle of contact is very small or zero, an example being very clean glass and water. In this case the liquid is said to 'wet' the solid.

(iii) $\theta > 90°$. Now we say that the liquid 'does not wet' the solid, the best example being mercury and clean glass where θ may be about 136°.

There are various ways of measuring the 'wettability' of solids by liquids which are, in essence, the same as specifying the angle of contact. This angle is not always a constant but depends on whether the liquid is advancing or receding over the solid surface. These two angles are often known as advancing and receding angles of contact and θ_A is usually larger than θ_R (Fig. 9.10). To

Advancing   Receding

Fig. 9.10 (a) Advancing and (b) Receding angles of contact.

determine the angle of contact of a liquid against a particular solid, a plane of the solid is tilted until the liquid surface is completely horizontal (as tested optically) right up to the solid's surface. This is illustrated in Fig. 9.11.

Fig. 9.11 Measuring the angle of contact.

9.5 SPREADING

Consider two liquids A and B whose free energies are respectively γ_A and γ_B where these are the values of γ in contact with the vapour phases. If we take a column of liquid A of unit cross-section and assume that this column can be separated across a plane cross-section as in Fig. 9.12, then two new surfaces of unit area are created. Thus there is an increase of free surface energy of $2\gamma_A$, and $W_A = 2\gamma_A$ is called the **work of cohesion** of liquid A. In the same way $W_B = 2\gamma_B$ is defined similarly for liquid B.

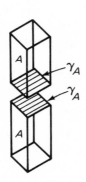
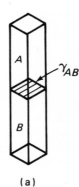
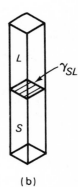

Fig. 9.12 Illustrating 'work of cohesion'.   Fig. 9.13 Illustrating 'work of adhesion'.

Suppose further that the two liquids A and B do not mix and consider a column of unit cross-section of the two liquids in contact as in Fig. 9.13a. Liquid A rests on liquid B and there is a plane interface between them. If we assume that the two liquids can be separated at this interface, the work done equals the change in free surface energy. This is $W_{AB} = \gamma_A + \gamma_B - \gamma_{AB}$ because two new unit surfaces of energies γ_A and γ_B have been produced where previously there was an AB interface of free surface energy γ_{AB}. W_{AB} is called the **work of adhesion**. Similarly if a liquid L is in contact with a plane solid surface S, then $W_{SL} = \gamma_S + \gamma_L - \gamma_{SL}$ is the work of adhesion (Fig. 9.13b).

Values of the work of adhesion for various organic substances in contact with water are given in Table 9.1 below; they are taken from a Unilever educational booklet *Surface Activity*, page 7.

Table 9.1

Substance	$\gamma_{AB} \times 10^3$ (Jm^{-2})
Paraffins	36 – 48
Aromatic hydrocarbons	63 – 67
Alkyl halides	66 – 81
Esters	73 – 78
Ketones	85 – 90
Nitriles	85 – 90
Primary alcohols	92 – 97
Fatty acids	90 – 100

Now consider the problem of a liquid spreading (a) on a plane horizontal solid surface and (b) on another liquid with which it does not mix.

Take first the surface of a solid. The reason why such a surface has a free surface energy γ_{SV} is because the molecules in it do not have a symmetrical environment; on one side of the surface there are neighbouring molecules of the solid, on the other only a few vapour molecules. If this solid surface is covered now by a liquid then this largely redresses the imbalance in the environment of the surface molecules. There are now comparable concentrations of neighbouring molecules both above and below the actual surface and so γ_{SV} is reduced to γ_{SL}.

If the liquid spreads over the solid's surface the liquid-vapour interface increases and so does the liquid-solid interface, whilst the solid-vapour one decreases. These changes cause the various free surface energies to change and spreading will proceed until the sum of the free energies is a minimum. Thus spreading occurs, provided that the increase in the free surface energy accompanying the area increase in the LV and LS interfaces is less than the decrease in energy due to the area decrease in the SV interface, that is if

$$\gamma_{LV} + \gamma_{LS} < \gamma_{SV} .$$

It is assumed in all this that the spreading film is thick enough for us to regard γ_{SL} and γ_{LV} as being independent. In a film which was only a few molecules in thickness one would have to look into this assumption in more detail.

Next consider two immiscible liquids A and B. If a drop of A is introduced on to the plane surface of B then the argument is the same as that for spreading over a solid surface; spreading now occurs if

$$\gamma_A + \gamma_{AB} < \gamma_B . \tag{9.6}$$

To sum up, if spreading of a liquid occurs either over a solid or over a second liquid surface, the spreading liquid moves so as to reduce the sum of the free surface energies.

If we return to the case of two liquids, we can define a spreading co-efficient S as

$$S = \gamma_B - \gamma_A - \gamma_{AB} .$$

Hence, from (9.6), spreading occurs if $S > 0$. Also, because $W_{AB} = \gamma_A + \gamma_B - \gamma_{AB}$ and $W_A = 2\gamma_A$, then $S = W_{AB} - W_A$. Thus spreading occurs if $W_{AB} > W_A$, that is if the work of adhesion for the two liquids is greater than the work of cohesion for the spreading liquid.

The equations

$$S = \gamma_B - \gamma_A - \gamma_{AB}$$

and $$S = \gamma_{SV} - \gamma_{LV} - \gamma_{LS}$$

really express the spreading behaviour of a so-called 'duplex' film. Such a film is of sufficient thickness for γ_A and γ_{AB} (or γ_{LV} and γ_{LS}) to act independently, that is the γ's for the upper and lower surfaces of the spreading liquid are inde-pendent. Duplex films are not thermodynamically stable and usually change into monomolecular films (monolayers). Indeed after the initial spreading as a duplex film, the molecules in the spreading liquid will often rearrange themselves into liquid 'lenses' in equilibrium with the monolayer. An example of this is benzene on water.

A monolayer behaves quite differently from both the substrate below it and the bulk of the liquid from which it is composed. In the spreading of monolayers the minimising of the total free surface energy is again the criterion. Monolayers are discussed further in Chapter 9.12.

9.6 INTRINSIC PRESSURE

Laplace showed that if one assumed that the effective range of inter-molecular forces is very short the surface energy of a liquid could be related to the latent heat of evaporation. This is done by comparing the effects of two different processes, the removal of a single molecule from the bulk of the liquid to a point a great distance away and the tearing apart of a layer of liquid from a bulk volume so that two new liquid surfaces are created. In its original form the argument simply gave an estimate of the effective *range* of the intermolecular forces in terms of the ratio of surface tension and latent heat of evaporation. However, if we make a definite assumption about the intermolecular forces (see, for example, Temperley *Properties of Matter,* p. 260 ff.) we can both obtain reasonably reliable estimates of the latent heat and surface tension and also relate them to the constant a in the van der Waals equation of state. In that theory a/V^2, the so-called **intrinsic pressure**, is closely related to the attractive forces between the molecules.

Consider a molecule B, part of whose sphere of influence of radius a lies above the surface as shown in Fig. 9.14. The resultant force on B due to its neighbouring molecules will then be directed inward ($\downarrow$) along the normal to the surface. This is because there are a greater number of molecules in the hemisphere below CD than in the volume CDD'C'C above CD; the nearer B is to the surface XX' the greater the inward resultant force on B. As the distance y of a molecule below XX' increases from zero to a, this inward force on B changes from its maximum value to its minimum value of zero and remains zero for all $y > a$ as for molecules like A in Fig. 9.14.

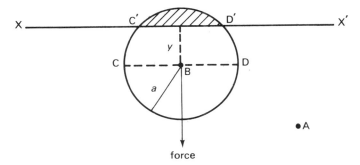

Fig. 9.14 The sphere of influence of a molecule below a plane surface.

Now consider a thin layer of liquid parallel to the surface whose upper and lower faces are at distances y_1 and y_2 below XX'. Both y_1 and y_2 are less than a (Fig. 9.15). The downward forces on each molecule in LL' are greater than those on each molecule in MM'. Thus because the layer is in equilibrium it

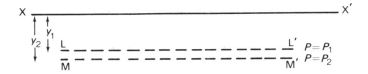

Fig. 9.15 The concept of **intrinsic pressure**.

follows that the pressure $P = P_1$ exerted on LL' by the liquid above the layer is less than the pressure $P = P_2$ exerted on MM' by the liquid below the layer. This result is valid for every such layer of liquid within a distance a of the surface, and the pressure P increases with distance y from the surface reaching a maximum at $y = a$. For all $y > a$ it remains constant. This pressure P which we have introduced is completely independent of gravitational forces and is called the **intrinsic pressure**. It is quite different from the hydrostatic pressure, which on the contrary is due to gravity. Note that in all this we have assumed that the concentration of molecules in the vapour phase above XX' is practically zero.

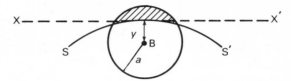

Fig. 9.16 The sphere of influence of a molecule below a curved surface.

Next consider a molecule B below a curved liquid surface SS' concave towards the bulk of the liquid as in Fig. 9.16. Let us compare the downward force on B with what it would be if the liquid surface were the plane surface XX'. In the case of the curved surface there is a greater volume of the sphere of influence above the liquid surface and so, from our reasoning above, there is a greater downward force on B in this case. Hence, by arguments similar to those above, it follows that the intrinsic pressure at a distance y below the upper point of the curved surface is greater than that at the same depth below a plane surface. Further, the pressure at this point below a curved surface increases as y increases to $y = a$ and is constant for all $y > a$.

Similarly when the liquid surface is convex towards the liquid, the pressure below the surface is *less* than that below a plane surface.

9.7 PRESSURE DIFFERENCE ACROSS A CURVED SURFACE

If a liquid is at rest its surface is generally flat at regions well away from solid boundaries. Near such boundaries it is in general curved. Surface effects can, as we have already mentioned and shall discuss again later in this Chapter, be represented by surface tension or surface energy. In consequence there is a pressure-difference across any curved portion of the surface.

To fix our ideas, suppose that we have a spherical bubble of radius r in equilibrium within the bulk liquid. We can determine this pressure difference $P_g - P_\varrho$ by using the principle of virtual work. If the bubble changes in radius by dr, the volume change is $4\pi r^2$ dr and the work done by the expanding gas is P_g times this, while the work done in pushing away the same volume of liquid is P_ϱ times this volume change. The difference between these two must be the work done in expanding the surface. If the expansion is supposed to take place adiabatically this must be $\gamma \, 8\pi r$ dr, so that we have

$$P_g - P_\varrho = \frac{2\gamma}{r} \, .$$

For more complicated geometry of a surface, it is always possible to find two principle radii of curvature for any surface element. The principle of virtual work can still be used to derive the pressure-difference $\gamma\left[\dfrac{1}{R_1} + \dfrac{1}{R_2}\right]$ if R_1 and R_2 are the principal radii of curvature.

9.8   CAPILLARY RISE

It is a well known experimental fact that a liquid for which the angle of contact θ is less than $90°$ will climb up a capillary tube immersed in it. For a liquid like mercury where θ is greater than $90°$, the liquid climbs down the tube instead.

Let us consider the capillary rise in a vertical capillary dipping into the free surface of a liquid of the former kind (Fig. 9.17). The meniscus is symmetrical

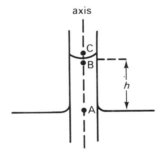

Fig. 9.17 Capillary Rise.

with respect to the axis of the tube. At its lowest point, which lies on this axis, the two principal radii of curvature will have the same value R (say). Consider the three points A, B and C as shown and let h be the height of the lowest point of the meniscus above the free liquid surface. The pressure P_c at C, just above the meniscus, is atmospheric. At B just below it is $2\gamma/R$ less. Thus

$$P_C = P_B + \frac{2\gamma}{R}.$$

At A, which is at the same horizontal level as the outer free surface of the liquid, the pressure P_A is also atmospheric and

$$P_A = P_B + \rho gh$$

where ρ is the density of the liquid. Since $P_A = P_C$ it follows that

$$P_B + \rho g h = P_B + \frac{2\gamma}{R}$$

or $$h = \frac{2\gamma}{\rho g R}. \tag{9.7}$$

Equation (9.7) is always valid and gives the capillary rise h in terms of R. The value of R is not readily measured and we must therefore make some assumption about the shape of the meniscus. It is usual to regard it as a spherical cap with an angle of contact θ (see Fig. 9.18).

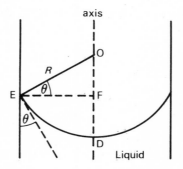

Fig. 9.18 The meniscus as a spherical cap.

If D is the lowest point of the meniscus and if E is the point in the plane of the diagram where the liquid meets the tube, then OD = OE = R where R is the radius of the spherical cap comprising the meniscus. If r is the bore of the capillary then

$$r = EF = R\cos\theta.$$

Substituting in (9.7) and eliminating R we have

$$h = \frac{2\gamma}{\rho g r} \cdot \cos\theta. \tag{9.8}$$

If θ is less than $90°$, as shown, $\cos\theta$ is positive and h is positive; that is, the liquid climbs up the tube. If θ is greater than $90°$, $\cos\theta$ is negative, so is h, and the liquid climbs down the tube (Fig. 9.19).

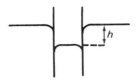

Fig. 9.19 Depression of mercury in a capillary.

If the liquid wets the tube then $\theta = 0$, the spherical cap becomes a hemisphere and

$$h = \frac{2\gamma}{\rho g r}. \tag{9.9}$$

The assumption that the meniscus is part of a sphere is of course an approximation. Indeed the calculation of the true shape of a capillary meniscus is very involved, although numerical tables compiled by Bashforth and Adams enable this to be done in certain cases.

9.9   SURFACE ENERGY, SURFACE CURVATURE AND VAPOUR PRESSURE

Consider a liquid in equilibrium with its saturated vapour. Such an equilibrium is really a dynamic one in which two effects are balanced, the rate of escape of molecules by evaporation from the free surface of the liquid and the reverse process in which molecules return to the liquid from the space above it. The saturation vapour pressure will depend on the number of molecules per unit volume in the space above the liquid. A molecule will leave the surface when it possesses a sufficient kinetic energy to overcome the attractions of its neighbours. Any factor which reduces the ability of a molecule to escape reduces the vapour pressure, and vice versa. One such factor is temperature since a lower temperature reduces the kinetic energy and a higher one increases it. Another factor, which is our immediate concern, is the curvature of the surface.

Fig. 9.20 Sphere of influence of a molecule in various surfaces.

Let us refer to Fig. 9.20. If a molecule escapes from the trough of a concave surface there are more molecules in the liquid to impede its departure than in the case of a plane surface. Consequently we expect the vapour pressure over a concave surface to be less than that over a plane one. Similarly a molecule escaping from the top of a convex surface suffers less hindrance from its neighbours in the liquid than occurs in the case of a plane surface; thus we expect the vapour pressure over a convex surface to be greater than that over a plane surface.

Consider a liquid in an enclosure containing only its saturated vapour and suppose that a vertical capillary is immersed in the liquid (Fig. 9.21). In this arrangement we have in the capillary a concave surface over which the S.V.P.

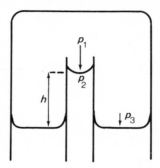

Fig. 9.21 Pressure over a plane and concave surface.

is P_1 and a plane one over which it is P_3. Let P_2 be the pressure in the liquid just below the meniscus, r the radius of curvature of the meniscus and h the height of the liquid in the tube. If ρ and σ are the densities of the liquid and the saturated vapour respectively, we have the equations

$$P_3 = P_1 + \sigma g h \quad \text{(vapour)}$$

$$P_3 = P_2 + \rho g h \quad \text{(liquid)}$$

$$P_1 - P_2 = \frac{2\gamma}{r} \ .$$

Eliminating h and P_2, we get for the difference between the pressure over the plane and concave surfaces

$$\Delta P = P_3 - P_1 = \frac{2\gamma}{r} \left(\frac{\sigma}{\rho - \sigma} \right), \tag{9.10}$$

showing that the pressure over a concave surface is less than that over a plane surface by this amount ΔP. Similarly, if we were to consider a liquid with an obtuse angle of contact which climbs down the tube, we could prove that the vapour pressure over a convex surface exceeds that over a plane one by an amount $\dfrac{2\gamma}{r} \left(\dfrac{\sigma}{\rho - \sigma} \right)$. When long columns are involved, that is when h is very large, equation (9.10) must be amended to take into account the compressibilities of the gas and liquid.

It is worth mentioning some work by La Mer and Gruen to verify (9.10) in a direct manner by balancing the increase of vapour pressure due to the curvature of a drop of solution against the decrease of the vapour pressure due to the presence of a dissolved solid. Further details may be obtained in the original paper.

We can discuss the evaporation of drops by considering a liquid in equilibrium with its saturated vapour in the space above it. To be more precise we should say that an equilibrium exists between the *plane* surface of the liquid and the saturated vapour. Now consider what happens to a drop of the liquid introduced into this space. If the drop is small enough it will start evaporating and energy must be supplied to produce the latent heat of vaporisation. Some of this energy is obtained from the loss of surface energy occurring as the radius of the drop decreases. Eventually when the radius of the drop decreases to a certain critical value the loss of surface energy is just enough to produce all the latent heat necessary for evaporation. All this implies the converse also, that a drop must have a nucleus containing several molecules before condensation can proceed. A simple way of estimating the 'critical' radius of a drop is as follows: the surface energy of a drop of radius r_c is $4\pi\gamma r_c^2$ while the energy needed to evaporate the drop is $4L\rho\pi r_c^3/3$, where L is the latent heat per unit mass. This gives $r_c = 3\gamma/\rho L$ which, in the case of water, leads to $r_c = 1.2 \times 10^{-8}$ cm.

9.10 METHODS OF MEASURING THE SURFACE ENERGY OF LIQUID-VAPOUR INTERFACES

In all this work the importance of absolutely clean conditions cannot be over-emphasized. For example, the presence of the minutest quantity of grease or oil from the fingers can form a monolayer which can have a marked effect on the surface energy. For such reasons the need for scrupulous cleanliness in both liquid and apparatus is important. Much of the earlier work is unreliable because of this.

Perhaps the most direct method of measuring surface energy is that involving capillary rise, discussed in Chapter 9.8. Here the only measurements necessary are the rise and the radius of the tube (see equation (9.9)).

Accurate work, using any of the standard methods, is very difficult. The interested reader is referred to specialised laboratory manuals for an account of the methods used and the precautions to be taken.

9.11 THE VARIATION OF FREE SURFACE ENERGY WITH TEMPERATURE

It is found that the free surface energies of all liquids decrease as the temperature is raised and become zero at the critical temperature, T_c. The decrease of γ is fairly linear over a large temperature range.

Many empirical relations between γ and the temperature, T, have been suggested. A well known one, due to Katayama, is

$$\gamma \left[\frac{M}{\rho_\ell - \rho_v} \right]^{2/3} = kT_c \left(1 - T/T_c \right)$$

where M is the molecular weight and ρ_ℓ, ρ_v are the densities of the liquid and vapour respectively. This equation agrees well with experiment even at temperatures near the critical point. Van der Waals had previously proposed the equation

$$\gamma = KT_c^{1/3} P_c^{2/3} \left[1 - \frac{T}{T_c} \right]^n$$

where K is a universal constant for liquids and n is a pure number; its value is about 1.21 for many liquids.

If the factor $\left[1 - \dfrac{T}{T_c} \right]$ is eliminated by means of the two previous equations, one obtains an equation first proposed by Macleod, namely

$$\gamma / (\rho_\ell - \rho_v)^4 = C$$

where C is a constant for a given liquid. If, following Sugden, we take the fourth root of C and multiply it by M, we obtain a molecular volume P' given by

$$P' = M\gamma^{1/4} / (\rho_\ell - \rho_v)$$

called the **parachor**. P' is independent of the temperature, and if γ is put equal to some value (unity, for convenience) then we have a method of comparing liquids under conditions of equivalent surface energy. The parachor was originally used to help to decide between different possible chemical structures of molecules but more sophisticated methods (e.g. infra-red spectroscopy) are now used.

9.12 ADSORPTION ON LIQUID SURFACES

Some substances such as grease will spread over the surface of a liquid to form a unimolecular film, that is a film just one molecule thick. Such a monolayer is called an **adsorbed** film. Much of the earliest work on such films was done by Pockels and Rayleigh; Langmuir later devised a method for studying these films in a more quantitative way. Langmuir's method was essentially as follows. A rectangular trough containing water is used in conjunction with two barriers A and B (Fig. 9.22). The movable barrier A rests on the sides of the trough and just touches the water surface. The barrier B floats on the surface of the water and its ends are joined to the sides of the trough by light ligaments of

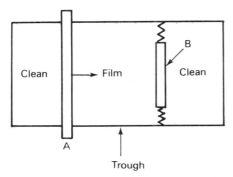

Fig. 9.22 Apparatus for studying a surface film.

metal foil. In an experiment the surface of the water is first swept clean by a number of movable barriers (type A). Then a very small amount of grease, dissolved in a solvent such as benzene which evaporates readily, is introduced on to the clean surfaces to the left of B. The barrier A is moved slowly to the right, thus compressing the surface film and causing a force to be exerted on B. (The metal foil prevents escape of the grease past the ends of B). The total force F pushing B to the right is measured by a torsion balance arrangement. Thus the surface pressure π, which is the force per unit length on B, can be calculated. Since the force on B is to the right, it implies that the surface energy is diminished by the surface film and we can write $\pi = \gamma_L - \gamma_{LF}$ where γ_L and γ_{LF} are respectively the surface tensions of the clean and contaminated regions. With this method it is possible to obtain curves relating the surface pressure π and the surface area σ of film per molecule. We can interpret such curves by saying that the observed surface pressure on B is a two-dimensional analogue of the pressure of a gas. This surface pressure can be thought of as being due to the impact of the stearic acid molecules on B. For large values of σ, the surface pressure is small and it increases slowly as the area of the film is decreased until σ is about 20 Å^2/mol. At this stage the surface film molecules form a tightly-packed two-dimensional arrangement and much larger values of π are required to produce further compression of the film.

Further studies of such films suggest that the long chains of stearic acid set in a vertical direction with their carboxyl 'heads' in water and the hydrocarbon chains standing out from the water surface (Fig. 9.23). The molecular volume of stearic acid is 556 Å^3, and if we take the value of 20 Å^2 as representing the cross-sectional area of a hydrocarbon chain this gives the length

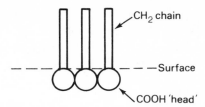

Fig. 9.23 Stearic acid molecules in a water surface.

of the stearic acid molecule as $556/20 = 27.1$ Å. This value agrees well with that obtained from X-ray diffraction measurements.

The molecules of an adsorbed monolayer can be regarded as a two-dimensional gas when the molecules are sparsely spread on the liquid surface. In fact the equation of state for a perfect gas in two dimensions can be written as $\pi\sigma = kT$, the counterpart of the more usual equation $PV = kT$. When more adsorbed molecules are present, these closer packed arrangements on the surface can be regarded as a two-dimensional liquid or solid.

It is found that certain chemical groupings, namely,

$$
\begin{array}{lll}
\text{carbonyl} & = & CO \\
\text{nitrile} & - & CN \\
\text{hydroxyl} & - & CH_2OH \\
\text{carboxyl} & - & COOH \\
\end{array}
$$

provide strong adhesion for spreading on water. These groups are polar and are called **hydrophilic**; this is the part of the molecule that forms the 'head' of the molecule directed towards the water in Fig. 9.23. The remaining part of the molecule is a hydrocarbon 'chain' which is non-polar and is **hydrophobic**; this is the upper part of the molecule in Fig. 9.23 and is directed away from the water.

One practical application of monolayer films has been used in water conservation, to slow down the rate of evaporation of water in reservoirs. For this purpose cetyl alcohol is used, 1 oz of which is sufficient to cover 3 acres of water with a monolayer. However, because weather effects cause a breaking up of the monolayer this must be replenished, and it is found that about 1 oz per acre per day is required to keep the monolayer in being over the whole surface. In Australia it is estimated that about 1 million gallons of water per acre per year could be saved by such means.

Another example of hydrophilic-hydrophobic properties is that of detergency. When dirt in the form of grease is to be removed from a fabric, the effect of the detergent solution is to cause an increase in the angle of contact of the grease with the material. As this angle of contact increases the grease becomes more and more globular in shape and finally breaks away as a ball from the material.

9.13 COLLOIDS

Colloids are substances which, though they appear to go into solution, remain dispersed as an assembly of discrete particles in the surrounding medium. A colloidal system consists of two phases, the continuous phase and the disperse phase. The continuous phase is the medium in which the colloid particles — the disperse phase — are dispersed. The continuous phase can be either solid, liquid or gas; so also can the disperse phase. In what follows we shall confine our attention to the case where the continuous phase is a liquid.

All colloids have a large surface-to-weight ratio and this explains the important role of surface properties in the behaviour of colloids. If the disperse phase is a solid we can have either a **sol** or a **gel**. To understand these terms we must remember that colloids tend to coagulate and precipitate if left alone for long enough. If this does not occur then one of two reasons will account for it;

(i)   There is a strong attraction between the colloid particles and the continuous phase in which they are dispersed. This case is called a **lyophilic sol**.

(ii) There is a strong repulsion between the various colloid particles which opposes their coalescence. This constitutes a **lyophobic sol**.

In certain cases a lyophilic sol can set as a solid containing both the disperse and continuous phases; this is known as a gel. In some instances we have reversible sol-gel systems known as thixotropes. Such a system is normally a solid (gel) but if shaken it behaves like a liquid (sol) and then resets as a solid on being left to itself. Thixotropic paints, mentioned in Chapter 13, illustrate this behaviour.

If the disperse phase is a liquid then we have an **emulsion**. An example is oil and water. Whether the arrangement is an emulsion of oil-in-water or water-in-oil depends on the volume ratio of the the two phases, and the method of preparation. The very creation of the suspension causes a large increase in the common surface area between the two liquids, and stability is reached when the interfacial surface energy tends to the lowest possible value.

9.14 CONCLUSION

Historically, in discussing the surface of a liquid the term which has been most used is that of 'surface tension'. Over the years there have been arguments as to the reality of such a tension. For example, an eminent authority such as Adam (1938) talks about a 'hypothetical' tension. Again, Champion and Davy (1936) describe surface tension as a 'useful fiction'. In recent books, such as *Properties of Matter* by Flowers and Mendoza (1970), the term 'surface tension' is again defined and used. Indeed, in the present Chapter we have seen that it is well-nigh impossible to explain certain phenomena without invoking the idea of a tension parallel to the liquid surface, for example the hanging liquid drop in Fig. 9.5. In certain simple experiments (e.g. with soap films on a frame or the movement of camphor chips on water) a force can really be measured, but this does not absolutely prove its existence on the surface of a *pure* liquid. In recent research work the objective existence of surface tension has been confirmed by Ono and Kondo (1960) and Shoemaker, Paul and de Chazal (1970); furthermore its value has been calculated from a knowledge of the intermolecular forces. This matter has been further discussed by Berry (1971) and by Walton (1969, 1972). We will therefore try to clarify the situation.

There is a good deal of evidence to show that, as might be expected, the surface layer of a liquid is depleted, that is the number density ρ of the molecules is less than that in the bulk liquid. Experiments using reflected polarized light suggest that this decreased number density is confined to a layer whose thickness is just a few molecular diameters.

We first consider Berry's explanation of a tension parallel to the liquid surface. If we consider a small imaginary plane area (a 'test surface') at a point inside a fluid in equilibrium, then the pressure at that point may be defined as the average normal force per unit area exerted by all the molecules on one side of the surface on all those on the other. Berry regards this pressure as being made up of two contributions. One part P_k is due to the transport of momentum by molecules (as in a gas) and its value is $P_k = \rho k T$ where ρ is the number density of molecules at the region in question. P_k is always positive. The second contribution P_f is due to the time average of the interactions between molecules on opposite sides of the test surface (c.f. the term a/V^2 in van der Waals's equation) and is particularly relevant when discussing a liquid or dense gas. We then write

$$P = P_k + P_f .$$

P_k is much greater for a liquid than for a gas. If the externally applied pressure is not too large, the molecules are not closely packed and so the attractive forces between the molecules dominate the repulsive forces. In this case P_f is negative in order to reduce the total pressure P to the value of that applied externally. If the external pressure increases, the molecules are squashed together and the repulsive forces predominate and thus further compression is resisted, P_f is then positive.

Let us refer to Fig. 9.24 which is taken from Berry's paper. In the bulk vapour (A) or bulk liquid (C) there is directional symmetry in the distribution of the molecules and so the pressure has the same value P_0 irrespective of the orientation of the test surface used to define it. However, in the surface region (B), the tangential pressure P^t and the normal pressure P^n will not be the same because there is no longer symmetry of direction, there being depletion in the

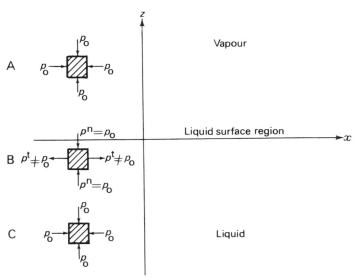

Fig. 9.24 Berry's theory.

vertical z direction but not in the horizontal x direction. However, since the fluid is in equilibrium the forces on the opposite faces of the three small cubes shown must be equal and opposite. Thus, neglecting gravity changes between A and C, the normal pressure P^n has the same constant value P_0 between A and C, right through the surface layer. On the other hand, the tangential pressure P^t is equal to P_0 in the bulk vapour at A and in the bulk liquid at C, but is *not* equal to P_0 in the surface layer.

Berry then proceeds to explain how the two contributions P_k^t, P_f^t to P^t vary with z. We shall not give the argument here but refer the interested reader to the original paper. Berry concludes that P^t varies with z in the manner shown by the dotted curve in Fig. 9.25. As we see, this curve goes negative showing that

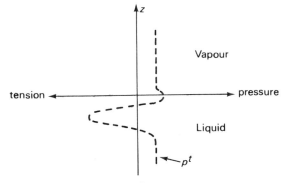

Fig. 9.25 Variation of P^t across a liquid surface.

there is a tangential tension near the liquid surface. Berry concludes that 'the surface layer of liquid, in contrast with the bulk, must possess rigidity in order to resist the shear stress that results from P^t differing from P^n; this is the basis for the statement appearing in older textbooks that liquids behave as if their surfaces are covered by an "elastic skin"'.

In his paper Walton (1969) considers a section of film stretched between two wires A and B (Fig. 9.26). If B is fixed and A is free to move, then an external force F must be applied to A to maintain equilibrium. If we consider any section C of the film it follows that for equilibrium the portion CB of the

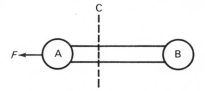

Fig. 9.26 Film of liquid stretched between two wires A and B.

film must exert on the portion AC a force equal and opposite to F. Does this tension in the film act all the way across the film thickness (as in a stretched sheet of solid) or does it act in the surface regions only? That is, is it a bulk tension or a surface tension? It is in fact a surface tension because F remains unchanged as the film thickness is reduced by evaporation. If the force F is increased in value and maintained, the film will expand until it breaks. Throughout this expansion the surface tension remains constant even though the film's surface area is increasing.

Walton shows that surface tension occurs as a consequence of assuming that the number of molecules which diffuse from the bulk of a liquid to its surface per second equals the number which leave the surface for the bulk. In his treatment he uses the cell model of a liquid in which each molecule is surrounded by about ten nearest neighbours. He argues that the average potential energy ϕ' of an enclosed molecule as it is displaced along a chosen direction x is as shown in Fig. 9.27. At A, the molecule is at its cell centre. To escape from

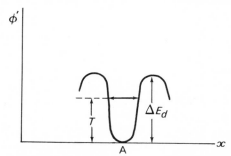

Fig. 9.27 Potential energy curve of an enclosed molecule.

the cell to the right in the x-direction it must acquire a total energy ΔE_d, the diffusion activation energy. When the kinetic energy T of the enclosed molecule exceeds ΔE_d, escape is possible and diffusive flow occurs. This is however a slow process; for most of the time the molecule is confined to its cell and only occasionally will T be large enough for it to escape. Walton states that in his two-dimensional liquid model (see Walton and Woodruff, 1969) the fraction of 'successful' attempts at escape is about one in a thousand at room temperature.

Because of the depletion in the surface layer, the average distance between neighbouring molecules is r', which is greater than the equilibrium separation r_0. Thus molecule A in Fig. 9.28 must now escape from the increased attractions

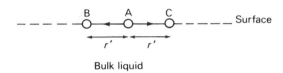

Fig. 9.28 Three molecules in a liquid surface.

of neighbours such as B and C before it can enter the bulk liquid. These forces acting parallel to the surface on a typical surface molecule like A therefore give rise to a 'surface tension' according to Walton's theory. By assuming an *LJD* form of $\phi(r)$ for liquid neon, Walton obtains a value of 0.5×10^{-3} Nm^{-1} compared with an experimental value of 5.5×10^{-3} Nm^{-1}.

Walton also uses similar diffusion arguments to explain capillarity. He considers how a liquid rises in a glass capillary tube which is lowered into the liquid. He assumes that a molecule of the liquid first requires an energy ΔE_e to escape upwards from the liquid, the latent heat of vaporization. If this molecule then settles on the glass wall of the capillary an amount of energy ΔE_a will be surrendered, the heat of adsorption. Thus to escape from the liquid to the glass a molecule must have a kinetic energy of at least ΔE_e; conversely a molecule which has settled in a site on the glass requires a kinetic energy of at least ΔE_a to return to the bulk liquid. If $\Delta E_a > \Delta E_e$ more molecules will diffuse up the glass wall from the bulk liquid than vice-versa, and **capillary rise** occurs. As a molecule A at the top perimeter of the meniscus diffuses up the glass wall it pulls on other neighbouring molecules. This explains the upward force around the perimeter of the tube at the top of the meniscus. As the rise of the liquid in the tube continues, further diffusion up the glass wall becomes increasingly difficult. Molecules like A at the wall at the top of the meniscus now experience the downward weight of the liquid column. This means that an energy greater than the initial ΔE_e is now required to escape from the bulk liquid. Eventually a state of dynamic equilbrium between these molecules diffusing up the glass from the liquid and those returning down the glass to the liquid will be reached. At this stage the weight contribution on A is sufficient to increase the value of ΔE_e to ΔE_a, and no further capillary rise occurs.

REFERENCES

Books
Surface Activity, Unilever Educational Booklet, 1968.
Adam, N. K., *The Physics and Chemistry of Surfaces,* Oxford, 1930.
Champion, F. C. and Davy, N., *Properties of Matter,* Blackie and Son, 1936.
Condon, E. U. and Odishaw, H., ed., *Handbook of Physics,* Part 5, Chapter 7, McGraw-Hill, 1958.
Feather, N., *An Introduction to the Physics of Mass, Length and Time,* Edinburgh University Press, 1959.
Flowers, B. H. and Mendoza, E., *Properties of Matter,* John Wiley and Sons Ltd., 1970.
Moore, W. J., 1963, *Physical Chemistry, 4th Edition,* Chapter 18, Longmans.
Sprackling, M. T., *The Mechanical Properties of Matter,* The English Universities Press Ltd., 1969.
Tabor, D., *Gases, Liquids and Solids,* Penguin Library of Physical Sciences, 1969.
Temperley, H. N. V., *Properties of Matter,* University Tutorial Press Ltd., 1963.
Zemansky, M. W. *Heat and Thermodynamics, 5th Edition,* McGraw-Hill Book Co., 1968.

Papers
Berry, M. V., 1971, *Phys. Ed.,* **6,** 79.
La Mer, V. K. and Gruen, R., 1952, *Trans. Faraday Soc.,* **48,** 410.
Ono, S. and Kondo, S., 1960, *Molecular Theory of Surface Tension in Liquids, Handbuch der Physik,* Vol X, Springer.
Shoemaker, P. D., Paul, D. and de Chazal, L. E., 1970, *J. Chem. Phys.,* **52,** 491.
Tabor, D., 1964, *Contemp. Phys.* **6,** 112.
Walton, A. J., 1969, *Contemp. Phys.,* **10,** 489.
Walton, A. J., 1972, *Phys. Ed.,* **8,** 491.
Walton, A. J. and Woodruff, A. G., 1969, *Contemp. Phys.,* **10,** 59.
Yarnold, G. D., 1960, *Contemp. Phys.,* **1,** 267.

Chapter 10
STUDIES OF LIQUID STRUCTURE

10.1 THE DISTRIBUTION FUNCTION $g(r)$

The two-molecule distribution function $g(r)$ has already been defined in Chapters 1 and 2.3. Knowledge of this function gives considerable information about the liquid. If the interaction function between two molecules is known (and if there are no more complicated interactions involving three or more molecules simultaneously) this determines the internal energy of the liquid as a function of pressure and temperature, from which all the other equilibrium properties can be obtained thermodynamically. If we also know the variations in $g(r)$ produced by gradients of temperature, density or stream velocity, we can in principle calculate the heat conductivity, diffusivity and viscosity respectively (see Chapter 12).

In this Chapter we shall review the available methods of determining $g(r)$ or predicting it theoretically. Some are referred to elsewhere in the book. Fuller accounts of most of them are given in *Physics of Simple Liquids* (1968) and we shall make appropriate references to this work.

We must distinguish carefully between studies on actual liquids and studies of models of liquids. The difficulty arises because it is known that even for molecules as simple as those of argon, the interactions are not strictly 'two-molecule'. In fact the effective interaction between two molecules can be modified by the proximity of a third. Therefore, if we are comparing theoretical and experimental results on the distribution function of liquid argon and find discrepancies, we are uncertain to what to attribute them. They may be due to the approximations made by theory, to inaccurate knowledge of the two-molecule interaction or to the effect of multi-molecule interactions.

Since the magnitudes of the latter are only roughly known and since in any case their effect cannot at present be assessed theoretically with any accuracy, a practice has grown up of comparing theoretical predictions with computer studies of various idealised models. The model most used for this purpose has been the 'rigid-sphere' model, for which the repulsion jumps from zero to infinity once the centres of two molecules are one diameter apart or less. Various refinements, involving the introduction of attractive forces of strictly 'two-molecule' type have also been made and enable more detailed checks of theory to be carried out.

10.2 THREE-DIMENSIONAL MODELS: BERNAL, SCOTT AND OTHERS

We have already referred briefly to Bernal's work in Chapter 4.6. It was concerned with the random packing of spheres and in one series of experiments he placed a number of plasticine spheres, chalked to prevent them sticking, into a rubber bladder. After removing the air to ensure that there would be no entrapped bubbles he compressed the spheres together so that they filled the whole of the space. On subsequent inspection the compressed spheres had become polyhedra, the average number of faces being 13.6. Furthermore the majority of the polyhedral faces were five-sided. Using the theory of Voronoi polyhedra, Bernal argued that in a heap of such polyhedra, having predominantly five-fold faces, there will be a preponderance of five-fold 'rings' of polyhedra. This must lead to irregular packings because such five-fold symmetry cannot lead to a *long-range* three-dimensional arrangement. Crystallography shows that if we have a spatially periodic lattice structure we can have 2-fold, 3-fold, 4-fold or 6-fold symmetry but not 5-fold. Bernal concludes that 'the inevitable appearance of five-fold rings eliminates any possibility of long-range order'.

In a second set of experiments Bernal and his colleagues studied the random packing of steel ball-bearings. Some 3000 quarter-inch bearings were packed into a rubber balloon and then squeezed and shaken so that random close packing of maximum density was achieved. Paint was poured in and the whole assembly left to dry, after which the balloon was removed. The assembly was then broken up and the number of contacts and near-contacts for 500 balls determined. The results gave coordination numbers varying between four and eleven and also enabled the radial distribution function to be obtained. This follows quite closely the *experimental* distribution function for liquid argon obtained from neutron diffraction work. All this shows that the radial distribution function for random close packing of the steel balls is essentially the radial distribution function of a simple monatomic liquid. It also shows that the *average* coordination number in a liquid is about 9 compared with 12 in a close-packed solid. If we consider the nearest-neighbour separation to be about the same in liquid and solid, this shows how the greater specific volume of the liquid can be explained by the decrease in coordination number. Bernal's work is essentially concerned with the geometric structure of a liquid and he refers to it as 'statistical geometry'.

Scott and his co-workers in Canada also carried out some experiments on dense random packing of steel ball-bearings. This 'dense' random packing, which means random close packing of maximum density, was obtained by tapping and shaking the vessel containing the balls. From a series of measurements it was possible to obtain the packing density d defined by

$$d = \frac{\text{Total volume of } N \text{ balls}}{\text{Volume of vessel containing } N \text{ balls}}$$

and the value of $d_1 = 0.637$ was obtained for the dense random packed assembly. We can also work out the value of d_2, the *maximum* value which occurs for a close-packed cubic or hexagonal crystal lattice for which $z = 12$. We find that $d_2 = 0.740$. From this it follows that

$$\frac{d_2}{d_1} = \frac{0.740}{0.637} = 1.16 .$$

If we regard the dense random packing d_1 as representing the liquid phase and the close-packed lattice d_2 as the solid, we can compare the above ratio with the ratio of the densities of solid and liquid argon at its melting point. This is just over 1.15, which agrees very closely with the above figure.

In some other experiments Scott used molten paraffin wax to 'bind' his assembly of balls in much the same way as Bernal used paint, and he thus obtained the radial distribution function. In Fig. 10.1 the radial distribution functions obtained from the random packing models of Bernal and Scott are compared with the experimental curve for liquid argon obtained from neutron scattering.

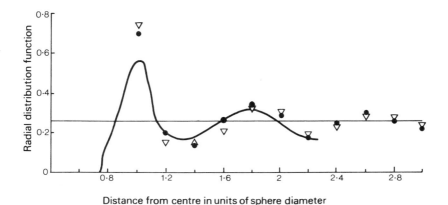

Distance from centre in units of sphere diameter

Fig. 10.1 Radial distribution function for liquid argon (from neutron scattering) plotted against distance from centre in units of sphere diameter; ($\triangledown$) from Scott's model; ($\bullet$) from Bernal's model.

10.3 TWO-DIMENSIONAL MODELS: WALTON AND WOODRUFF

The three-dimensional aggregate of painted steel balls in Bernal's model really represents a 'flash photograph' of the liquid. The same is true of Scott's model. It is a 'static' model and does not allow the effect of the thermal motion of the molecules to be examined. Furthermore no attempt is made to allow for attractive forces between the balls, though repulsive forces are present since we are dealing with 'rigid' spheres.

We now discuss the two-dimensional liquid-simulator devised by Walton and Woodruff previously mentioned in Chapter 4.4. This model consisted of ball-bearings on a rough-moulded glass tray. It is an improvement on the Bernal and Scott models in two ways: first the attractive force is provided by coating the ball-bearings with oil and secondly thermal agitation is simulated by vibrating the tray with an eccentric drive from a motor. They determined the force-separation curve for a pair of bearings by a balance method and found it to be of the general shape shown in Fig. 10.2, which is not very different from the force-separation curve between two real molecules shown in Fig. 3.1a.

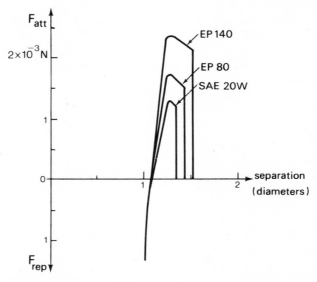

Fig. 10.2 The (force, separation) curve for a pair of bearings obtained by Walton and Woodruff using a balance method.

We have already described in Chapter 4.4 how the model shows a 'solid' melting into a 'liquid'. They also obtained a two-dimensional radial distribution function as follows. Choose the centre 0 of any bearing as origin. Then the number of bearings between distances of r and $r + dr$ from 0 may be written as $2\pi r\rho(r)\mathrm{d}r$ where $\rho(r)$ is the number density at radius r. For a three-dimensional arrangement the corresponding number is $4\pi r^2\rho(r)\mathrm{d}r$. The values of $\rho(r)$ can be found from simulator 'still' photographs by placing over the bearing a piece of transparent paper on which a series of concentric circles of centre 0 is drawn. This procedure is repeated about many bearings and the results are averaged. The radial distribution obtained shows the general 'peaks' and 'troughs' that one would expect, but no direct comparison with experiment is possible. Walton and Woodruff also attempted to simulate two-dimensional viscous behaviour by applying a constant couple to a short cylinder immersed in a tray of bearings with its axis of rotation perpendicular to the tray.

10.4 'COMPUTER EXPERIMENTS':
MONTE CARLO AND MOLECULAR DYNAMICS METHODS

These methods have been referred to in Chapter 3.6. The basic principles are simple, but their execution requires a large and fast computer. Even then we can only deal with 'assemblies' of up to a few hundred molecules, but the properties of these assemblies turn out to be not very different from those of larger ones. Thus by working with 'assemblies' of different sizes we can extrapolate the results to real assemblies containing very large numbers of molecules.

The Monte Carlo method starts off with the molecules in some standard arrangement. This is then progressively changed by altering the position of each molecule in turn, thus simulating the effect of thermal agitation. At various stages in the history of the 'assembly' we can calculate the quantities we are interested in, such as the mean kinetic energies per molecule or the molecular distribution function. The average of a large number of such samples taken at various stages in the history of the assembly can also give us an idea of how these quantities are changing with time, and this can be done in both equilibrium and non-equilibrium situations. By beginning with assemblies with non-uniform flow velocity, we can actually simulate processes like diffusion or viscous flow in a liquid. In other words, we can study transport problems almost as easily as we can equilibrium ones.

We are faced with two distinct problems. First we must devise a routine for progressively shifting the molecules, such that the successive configurations are an unbiassed or representative selection of those that might occur in a real assembly. Secondly, we must ensure that the process used for sampling configurations at various stages in the history really is 'fair' and unbiassed. This is the real difficulty of the Monte Carlo method; any slight flaw in the 'randomness' of the processes used can produce misleading results.

The molecular dynamics method is, in a sense, more ambitious. We study a small assembly by actually solving the individual equations of motion of all the molecules. The study of motions of even ten molecules in full detail requires a very large computer programme and is only possible for reasonably simple types of interaction. The processes of calculating average velocities, $g(r)$, energies and so on are then straightforward, and it is not necessary to devise any sampling process. Again we can study non-equilibrium processes almost as easily as we can equilibrium ones. In principle, it is only necessary to begin with the assembly having a non-uniform distribution of density or velocity and watch its evolution as time goes on.

For detailed accounts of these two methods, the reader is referred to *The Physics of Simple Liquids,* Chapters 4 and 5 (North Holland). One of the most significant discoveries was made independently by each of these two methods. This was that, at a density about 2/3 of close-packed density, a gas of rigid spheres would make a transition from disorder to order. This transition corresponded to the solidification transition of a real gas or liquid, and would still occur even if there were no attractive forces between the molecules at all. The existence of such a transition for the rigid sphere gas had been predicted theoretically in 1941 by Kirkwood and Monroe, but the prediction had remained suspect because of the drastic approximations that had had to be made in the theory. Indeed many physicists were inclined to believe that attractive forces between the molecules were needed in order to produce a sharp transition like melting.

10.5 STUDIES ON REAL LIQUIDS

Virtually the only method of determining the structure of matter is to bombard it with other particles, or to impart energy to it in some other way. Because of the very complicated motion of the molecules in a liquid we could normally expect to obtain very little information merely by firing particles through it at various speeds. An incident particle would emerge having made several collisions with the liquid molecules, and it might be very difficult to infer anything about the structure from the velocities and directions of the scattered particles. However, there is one important exception to this rather sweeping statement. Quantum mechanics teaches that any stream of material particles of mass m and velocity v are associated with a wavelength $\lambda = h/mv$ (where h is Planck's constant) known as the de Broglie wavelength. Now, if we can work with material particles with an effective wavelength comparable with the mean distance between molecules in a liquid, which is of the order of 1-2 Angström units (= 10^{-10} metres), we might expect diffraction effects similar to those produced by X-rays, whose wavelength is also of this order. After the fundamental discovery, by von Laue and the Braggs, that we can determine the structure of crystals from their X-ray diffraction patterns, it was natural to turn to liquids. X-ray studies of liquids developed rapidly in the 1920's and 30's, once it was realised that a non-crystalline substance could still produce a diffraction pattern.

The requirement that the wavelength must be comparable with the spacing of liquid molecules imposes sharp limits on the types of energy and the types of material particle that can be used in this way. We consider the possibilities in turn.

(i) **Visible and ultra-violet light.** Wavelengths are of the order of thousands of Angströms, and it is obvious that they can give very little *direct* information about the micro-structure of a liquid. Near the critical point however, density fluctuations can extend over thousands of atomic distances and possibilities do exist for using laser light to get information about this. (See Chapter 11 of *Physics of Simple Liquids*).

(ii) **Ultrasonics.** The situation is similar to that of visible light. For example, in water the velocity of sound is about 1500m/sec. To get a wavelength of the order of Angströms would require an impossibly high frequency of the order of 10^{13} HZ. However, like optical studies, scattering of ultrasound has been valuable in throwing light on density fluctuations near the critical point. In addition, see Chapter 7, we can study the variation of compressibility and absorption as we approach the liquid-vapour transition. (See also Chapter 9 of *Physics of Simple Liquids*.) Such information is of considerable theoretical interest at the present time, since the critical region is being studied both theoretically and experimentally.

(iii) **X-rays and gamma rays.** These are the only electromagnetic radiations with wavelengths in the right range. Most of the studies have been carried out with X-rays since they are easier to produce, but some preliminary work has been done with gamma rays.

(iv) **Material particles.** We have, apparently, a very wide variety of choice, electrons, positive and negative ions, neutral particles. In fact a great deal of work on the passage of ions through liquids has been done, but it throws only indirect light on the micro-structure. This is because an ion carries with it an 'atmosphere' of liquid molecules as a result of its large electric field, the effect being roughly similar to the Debye-Hückel effect in electrolytes. Put in another way, the intrinsic structure of a liquid is always badly distorted by the passage of a charged particle.

It *is* possible, by special methods, to accelerate neutral atoms to high velocities by accelerating them as positive ions which then recombine with electrons, but it is far simpler to use neutrons from the points of view of both production and handling. Their velocity can be reduced to any desired extent by passing them through hydrogen-bearing materials, and it is possible to obtain extremely narrow and accurately known velocity distributions by time-of-flight and 'chopping' techniques. As we shall see below, we can theoretically obtain more information from neutrons than from X-rays, because we can measure the velocities as well as the intensity of scattered neutrons as a function of scattering angle.

If we are interested in the intrinsic equilibrium structure of a liquid such as argon, almost the only parameter of interest is the so-called distribution function $g(r)$, describing the way that the number density fluctuates in the neighbourhood of a particular molecule. In these circumstances virtually the only methods available are X-ray and neutron scattering. This rather sweeping statement needs qualifying. In a molecular liquid there is also the possibility of using a typical molecule of the liquid as a **probe**. We can, for example, use the shifts in the energy-levels of a molecule (detected by infra-red or ultra-violet spectroscopy) to give information on the change in environment on melting. If the molecule has a magnetic moment we can get information about the change in environment of a typical molecule by nuclear magnetic resonance — the magnetic moment of the nucleus is here used as a probe. Another very ingenious idea, due to Casanova and Levi (*Physics of Simple Liquids,* Chapter 8), was to compare the relative abundance of argon isotopes in liquid and vapour. There is a dependence of the entropy of each phase on the relative isotope abundance. None of these methods actually measure the distribution function itself but give information about various averages of it over distance and angle. However, such information can be valuable because the various averages depend mainly on the short-distance behaviour of the distribution function. This is just the region that is hardest to measure by scattering experiments, as we shall see below.

10.6 THE DISTRIBUTION FUNCTION OBTAINED FROM X-RAY AND NEUTRON SCATTERING

We summarise the fundamental relation between the intensity of scattered X-radiation and the distribution function of a liquid or gas. For a definitive account of the applications of this relation and of the corrections that are made in practice the reader is referred to Pings (Chapter 10 of *Physics of Simple Liquids*).

We take our origin of coordinates as the centre of one of the molecules and consider a beam of X-rays incident on it from the left, the direction of propagation being specified by the unit vector S_0. This beam is scattered by the electrons in the molecule at the origin, and is also scattered by the electrons in other molecules at various distances from this particular molecule. If we average over time, by definition the mean density of scattering electrons at a point distant r from our particular molecule is proportional to the distribution function $g(r)$.

In other words, we have interference between the X-rays scattered by our typical molecule and by the X-rays scattered by other molecules surrounding it at a distance r. If we examined the total radiation scattered out of the beam at an angle determined by the unit vector S_1 by just two molecules separated by a distance r, the path difference depends on both r and the angle. In Fig. 10.3 let

0 be the origin and P the position of a typical molecule and let OP be the vector *r*. Erecting perpendiculars to S_0 and S_1 at 0 and P respectively, the path

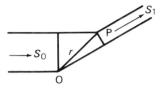

Fig. 10.3 Diffraction geometry.

difference for waves scattered from 0 and P is $\mathbf{r} \cdot (\mathbf{S}_1 - \mathbf{S}_0)$ and the corresponding phase shift is $2\pi\mathbf{r}\cdot(\mathbf{S}_1 - \mathbf{S}_0)/\lambda$ where λ is the wavelength. If the wave scattered from 0 is $A_0 e^{i\omega t}$, that scattered from P is $A_0 \exp\left[i\omega t + \dfrac{2\pi i}{\lambda}\mathbf{r}.(\mathbf{S}_1 - \mathbf{S}_0)\right]$.

If we now replace the single scattering centre at P by a continuous distribution of scattering centres $\rho(r)$ the resultant scattered amplitude becomes, integrating over the volume V of the liquid,

$$A_0 \exp(i\omega t) \int_V \rho(\mathbf{r}) \exp\left[\frac{2\pi i}{\lambda} (\mathbf{r}.(\mathbf{S}_1 - \mathbf{S}_0))\right] d\mathbf{r} \ .$$

We cannot measure the *amplitude* of X-rays directly, and one can only make inferences about *phase* by indirect methods, but *intensity* can be measured relatively easily, for example by a photographic plate or an ionisation chamber.

The intensity relative to that of a single scattering molecule is the square of the modulus of the above expression divided by A_0. We have a relation between $\rho(r)$ and the scattering intensity $I(S)$ as a function of angle or equivalently of S, the length of the vector $(\mathbf{S}_1 - \mathbf{S}_0)$. This relation can, by Fourier transform techniques, be inverted to give (to a factor)

$$r\rho(r) \propto \int_0^\infty S\,I(S) \sin (Sr)dS \ .$$

Thus, if we could measure $I(S)$ for all values of S, that is to say for all angles of scattering between 0 and $\pi/2$, we could in principle determine the scattering density $\rho(r)$ for all values of r. In practice $I(S)$ is difficult to measure for low values of S (small scattering angle) because of the presence of the unscattered beam, and this implies a corresponding uncertainty in $\rho(r)$ for *large r*. Equally, $I(S)$ is difficult to measure accurately for large values of S, because these correspond to values of scattering angle near $\pi/2$, and a small change in angle implies a very large change in S. The latter uncertainty is more serious because it means that $\rho(r)$ is uncertain for *small r,* which is where we want the most accurate information. For example the calculated pressure and internal energy of a liquid are both very sensitive to the actual form of $\rho(r)$ (and hence to the molecular distribution function $g(r)$) for small r where the potential energy of interaction between two molecules is large.

It remains to ask what additional information can be obtained from neutron scattering. The neutron beam of constant velocity can be represented as a de Broglie wave and we can define a density of scattering centres in a way very similar to what we have just done for X-rays, provided that we recognise that X-rays are scattered by electrons and neutrons by nuclei. For neutrons, we can measure not only intensity but also neutron velocity as a function of scattering angle. What additional information can we get from the changes in velocity?

We can look at this from two different points of view. First, we can define an experimentally measurable function, usually written $S(k,\omega)$, which describes the intensity of scattered neutrons as a function of both scattering angle and momentum. By an argument analogous to that used for X-rays we can deduce from this the function $\rho(\mathbf{r},t)$, which is the expected density of scattering nuclei at position $\mathbf{r}$ and time t given that there was a scattering nucleus at the origin at time $t = 0$ (see J. E. Enderby, *Physics of Simple Liquids,* Chapter 14). Thus we obtain information on the distribution function $g(r)$ as a function of t as well as of r, whereas X-ray experiments only give information on the *time-average* of $g(r)$. Another interpretation of $S(k,\omega)$ is that it gives information about the **elementary excitations** in liquids. According to the Debye picture the motion and interactions of all the molecules can be thought of as a distribution of sound quanta of various frequencies. If we wish, we can relate $S(k,\omega)$ to the exchange of momenta between neutrons and these elementary excitations, thus giving information about their time-history. We can also relate $S(k,\omega)$ to the time-history of $g(r)$. Clearly the variation of $g(r)$ with time is related to the transport properties of the liquid.

The reader is referred to Chapter 14 of *Physics of Simple Liquids* for a discussion of neutron scattering measurements and to Chapter 12 of the same book for discussions of the relations between the observed $S(k,\omega)$ and the time-correlations and transport properties of a liquid.

10.7 DISCUSSION

There is reasonable agreement between the various methods of measuring the distribution function of a liquid, the theoretical predictions based on integral equations and the approaches using mathematical models and computer experiments. However, a comparison of the distribution function of actual liquid argon (with attractive and repulsive forces between molecules) with the distribution function for a rigid sphere gas of comparable density also shows close agreement! In other words, the distribution function $g(r)$ is insensitive to the form of the interaction function $\phi(r)$. On the other hand, calculations of pressure and internal energy are sensitive to the form of $g(r)$ for small values of r, just the region that is hardest to study by scattering experiments.

Various attempts have been made to relate $\phi(r)$ not to $g(r)$ directly, but to the *departures* of $g(r)$ from the form appropriate for (for example) rigid spheres which can be regarded as known. This work promises well, but is in its early stages. It is also possible (see Chapter 2 of *Physics of Simple Liquids*) to regard the theoretical relations between $g(r)$ and $\phi(r)$ in this same light: that $\phi(r)$ determines the *departures* of $g(r)$ from the form appropriate to some standard model, such as the rigid sphere one. The theoretical relationships can be derived by the following process. We assume a small change in the form of the inter-action function $\phi(r)$ and try to estimate the corresponding change that would occur in $g(r)$. This leads to an integro-differential equation for $g(r)$ which, as shown by Rushbrooke, can be converted into an integral equation for $g(r)$ if certain simplifying assumptions are made. (See Chapter 2 of *Physics of Simple Liquids*).

REFERENCES

Papers

Bernal, J. D., 1959, *Nature*, **183**, 141.
Bernal, J. D., 1960, *Nature*, **185**, 68.
Bernal, J. D. and Mason, J., 1960, *Nature*, **188**, 910.
Bernal, J. D., Mason, J. and Knight, K. R., 1962, *Nature*, **194**, 957.
Bernal, J. D., 1964, *Proc. Roy. Soc.*, **A280**, 299.
Scott, G. D., 1960, *Nature*, **188**, 908.
Scott, G. D., 1962, *Nature*, **194**, 956.
Walton, A. J. and Woodruff, A. G., 1969, *Contemp. Phys.*, **10**, 59.

References from *Physics of Simple Liquids,* edited by H. N. V. Temperley, J. S. Rowlinson and G. S. Rushbrooke, North-Holland Publishing Company, Amsterdam, (1968).
Rushbrooke, G. S., Chapter 2.
Alder, B. J. and Hoover, W. G., Chapter 4.
Wood, W. W., Chapter 5.
Bernal, J. D. and King, S. V., Chapter 6.
Casanova, G. and Levi, A. Chapter 8.
Sette, D., Chapter 9.
Pings, C. J., Chapter 10.
McIntyre, D. and Sengers, J. V., Chapter 11.
Schofield, P., Chapter 13.
Enderby, J. E., Chapter 14.

Chapter 11
MIXTURES AND SOLUTIONS

11.1 INTRODUCTION

In this Chapter we shall be concerned with mixtures of two nonreacting substances which form a true solution. In such a binary mixture it is found that every part of the solution is like every other part, that is a true solution is a homogeneous phase, an example being sugar dissolved in water.

Since the present book is concerned with the liquid state, only three types of binary solutions will be considered:
(1) The solution of a liquid in a liquid, i.e. a liquid mixture,
(2) The solution of a gas in a liquid, and
(3) The solution of a solid in a liquid.
It is usual to refer to the substance that dissolves as the **solute** and the substance in which the dissolving occurs as the **solvent**. In the case of a solid-liquid mixture for example, it is clear that the solid is the solute and the liquid the solvent. However in liquid-liquid mixtures where the two liquids dissolve in each other in all proportions, it is not so easy to decide which is the solute and which the solvent. On the whole, therefore, we shall not use these two terms.

The solubility curve is determined by various physical factors. Most important is the extent, if any, to which ions are formed in solution. In certain solutions the solute molecules, for example common salt, break up into charged particles (ions) Na^+ and Cl^-, but we shall confine ourselves to solutions in which the molecules of solute and solvent remain intact. In such solutions solubilities are determined by the relative sizes and shapes of the two types of molecule and the relative strengths of the interactions between like and unlike molecules. Chemists often make a rough distinction between 'good' and 'poor' solvents, according to whether it is energetically more favourable for unlike or like pairs of molecules respectively to be near one another.

It usually happens that chemically similar molecules have very similar properties in solution. Furthermore a chemically homologous series of substances, for example the paraffins or alcohols, often show a smooth change in their properties in solution (and in their other physical properties) as the size of the molecule changes. However, chemical composition is not entirely a safe guide to what may happen. Substances as chemically similar as the salts of the alkaline earth metals or the various allotropic modifications of sulphur and phosphorus have widely different solubilities. Both physical, structural and chemical considerations play their parts in a proper understanding of solutions. In general it is true to say that two substances of similar chemical nature will 'tolerate' each other's presence and hence form a solution most readily. Conversely, substances of very different chemical nature will generally show little inclination to form solutions. In between these two extremes there will be a large number of intermediate possibilities. Some liquid-liquid mixtures illustrate these facts quite clearly. For example:

(1) Water and ethyl alcohol, which are chemically similar, form a solution in which the two liquids dissolve in all proportions. We say that the two substances are **completely miscible**.

(2) Water and mercury do not dissolve in each other at all; they are chemically very dissimilar and are **completely immiscible**.

(3) In the intermediate stage between the preceding two cases we have **partially miscible** liquid pairs such as water and ether. Water is capable of dissolving a small amount of ether to form a saturated solution of ether in water; ether will also dissolve a small amount of water to form a saturated solution of water in ether. Thus for a high proportion of either liquid a completely homogeneous solution is formed. If however the proportions of the two liquids are not within these saturation limits then two solutions will coexist as layers — one being a saturated solution of ether in water and the other of water in ether.

Two other factors which affect solubility are the temperature and, to a lesser extent generally, the pressure.

Another factor that really ought to be considered in any discussion of the solution process is that the structure of the liquid may be changed by dissolving another substance in it. If the solute molecules are smaller than those of the liquid the effect could well be, on the average, to draw the liquid molecules closer together. If they are equal in size or larger the liquid molecules are likely, on the average, to be driven further apart. In other words, a liquid accommodates itself to the solute molecules in a manner analogous to an elastic bag. This effect is difficult to allow for quantitatively, and until very recently it has been neglected altogether in the discussion of solutions.

If we neglect altogether this change in liquid structure as a result of the solution process we can still describe this process to a reasonable first approximation. We can regard the liquid as a quasi-lattice and picture the solution process as a replacement of some of the liquid molecules by molecules of solute. One can simplify the model further by neglecting all intermolecular interactions except those between nearest neighbours on the quasi-lattice.

11.2 THE CONDITIONS FOR THE EQUILIBRIUM BETWEEN A SOLUTION AND ITS VAPOUR PHASE

Consider a binary liquid mixture in equilibrium with its mixed vapour. We refer to this as a two-phase assembly consisting of the liquid (condensed) phase and the gaseous (vapour) phase. It is shown in books on statistical mechanics and physical chemistry that the condition for equilibrium at constant temperature and pressure in such a two-phase two-component assembly is that the chemical potential μ of each component must be the same in both phases (see, for example, *Physical Chemistry* by Moore and also Appendix 2.)

If there are M_A, M_B molecules present in the liquid phase of each of two liquids A and B and if A_l is the total Helmholtz free energy of the mixed liquid phase, then the corresponding chemical potentials are

$$\mu_{A_l} = \frac{\partial A_l}{\partial M_A} \quad , \quad \mu_{B_l} = \frac{\partial A_l}{\partial M_B} .$$

Similarly if there are N_A and N_B molecules of each kind in the mixed vapour phase, and if A_g is the total free energy of this phase, then

$$\mu_{A_l} = \frac{\partial A_g}{\partial N_A} \quad , \quad \mu_{B_l} = \frac{\partial A_g}{\partial N_B} .$$

In that case the equilibrium condition demands that

$$\mu_{A_l} = \mu_{A_g}$$

for the A component and

$$\mu_{B_l} = \mu_{B_g}$$

for the B component.

These conditions will be applied in Chapter 11.3 below.

11.3 LIQUID MIXTURES; PERFECT (IDEAL) SOLUTIONS

In considering a binary liquid mixture in equilibrium with its mixed vapour we shall use the approach considered by Rushbrooke (1949) in his book *Statistical Mechanics*. This is to consider the mixed liquid phase as a quasi-lattice in which a site is occupied by either an A or B molecule. In such a quasi-lattice the lattice sites can be considered as the mean positions in the liquid phase.

Let there be N_A, N_B molecules of each kind in the vapour phase, M_A, M_B

of each kind in the liquid phase, and let X_A, X_B be the total number of each kind. Hence, $X_A = M_A + N_A$, $X_B = M_B + N_B$. We assume that molecules of types A and B are of about the same size so that each type can occupy a similarly sized volume in our quasi-lattice. If this is so we can regard our quasi-lattice as comprising $(M_A + M_B)$ lattice sites and the total number of different arrangements obtained by permuting the molecules on all these sites is

$$\frac{(M_A + M_B)!}{M_A! \, M_B!} \, . \tag{11.1}$$

Using the methods of statistical mechanics the Helmholtz free energies A_l, A_g of the condensed and gaseous phases can be obtained. Hence the chemical potentials μ_{A_l}, μ_{A_g} for the A component in both phases can also be obtained and similarly μ_{B_l}, μ_{B_g} for the B component. Putting $\mu_{A_l} = \mu_{A_g}$ and $\mu_{B_l} = \mu_{B_g}$ then leads, after certain steps, to the results

$$P_A = \frac{M_A^*}{M_A^* + M_B^*} \, P_A^\circ \tag{11.2a}$$

and

$$P_B = \frac{M_B^*}{M_A^* + M_B^*} \, P_B^\circ \tag{11.2b}$$

where M_A^* and M_B^* are the *equilibrium values of* M_A and M_B.

In the above treatment three assumptions have been made:
(1) It is assumed that the vapour phase consists of a perfect mixed gas and that the volume of the condensed phase is negligible compared with that of the vapour phase.
(2) It is assumed that there are no preferential interactions between molecules of the same or different kind in the condensed phase. That is, an *AA, BB* or *AB* nearest-neighbour pair makes the same contribution to the total energy of the assembly.
(3) It is assumed that an A and B molecule have approximately the same size and shape.

We now return to equations (11.2a) and (11.2b). In these equations P_A, P_B are the partial vapour pressures of the two components in the gas phase; P_A°, P_B° represent the vapour pressures of the pure A and B components at the same temperature. These equations embody the properties of a **perfect** or **ideal** solution and are a statement of **Raoult's law**. This states that for a perfect solution the partial vapour pressure in the gas phase is proportional to the mole fraction in the condensed phase, that is in solution.

The two mole fractions are defined by

$$x_A = \frac{M_A}{M_A + M_B} \quad , \quad x_B = \frac{M_B}{M_A + M_B} \qquad (11.3)$$

and so
$$P_A = x_A P_A^{\circ} \qquad (11.4a)$$

$$P_B = (1 - x_A) P_B^{\circ} \qquad (11.4b)$$

since $x_A + x_B = 1$. Note that in equations (11.3) we have omitted the asterisks introduced in equations (11.2) and that M_A, M_B are from now on understood to be the equilibrium values. A plot of equations (11.4a) and (11.4b) at constant temperature is given in Fig. 11.1.

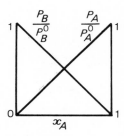

Fig. 11.1 Graphs illustrating Raoult's law (at constant temperature) for a perfect solution.

The gain in entropy on mixing the two pure liquids to form the perfect solution is given by

$$\Delta S = S_l - S_A - S_B$$

where S_A and S_B are the entropies of the two pure liquids making up the mixed condensed phase of entropy S_l. The value of ΔS, called the entropy of mixing, turns out to be

$$\Delta S = k \ln \frac{(M_A + M_B)!}{M_A! \, M_B!} . \qquad (11.5)$$

It is found that the corresponding change in free energy ΔA turns out to be negative (that is $\Delta A = -T \Delta S < 0$) and the change in internal energy, ΔU, is zero.

In practice very few solutions obey Raoult's law very closely and the experimental values of P_A/P_A° and P_B/P_B° lie either above or below the straight lines of

Fig. 11.1. These cases are referred to respectively as positive and negative deviations from Raoult's law. For the special case of Raoult's law and also for the cases where the graphs deviate from it, the two partial vapour pressure curves are related to each other and are not independent. This relation is given by

$$\frac{x_A}{P_A} \frac{\partial P_A}{\partial x} + \frac{x_B}{P_B} \frac{\partial P_B}{\partial x} = 0 \qquad (11.6)$$

called the Duhem-Margules equation. For further discussion see *Principles of Chemical Equilibrium* by Denbigh or *Physical Chemistry* by Moelwyn-Hughes. This equation holds whether x_A or x_B is inserted for x in this equation. The partial differentiation w.r.t. either x_A or x_B is performed at constant temperature.

Finally we note that the partition function Z for the condensed phase may be written as

$$Z = f_A^{M_A} f_B^{M_B} \frac{(M_A + M_B)!}{M_A! M_B!} \qquad (11.7)$$

where f_A and f_B respectively are the partition functions for an A and B molecule in this phase.

11.4 LIQUID MIXTURES; REGULAR SOLUTIONS

Let us now suppose that assumption (2) of the last section is no longer valid and that there *are* now preferential interactions between AA, BB and AB nearest-neighbour pairs. Let us further suppose that each molecule on our quasi-lattice has z nearest neighbours.

If each of the AA, BB and AB nearest neighbour pairs contributes W_{AA}, W_{BB} and W_{AB} to the total energy and if the numbers of such pairs are respectively M_{AA}, M_{BB} and M_{AB}, then that particular configuration makes a contribution to the total energy of the assembly of

$$U_c = M_{AA} W_{AA} + M_{BB} W_{BB} + M_{AB} W_{AB} . \qquad (11.8)$$

If we count all the nearest neighbours of all the M_A A-molecules we get zM_A and we may write

$$zM_A = 2M_{AA} + M_{AB}$$

since each AA pair is counted twice in the total zM_A. Similarly

$$zM_B = 2M_{BB} + M_{AB} .$$

Using these last two equations and substituting·in (11.8) we get

$$U_c = \tfrac{1}{2} z M_A W_{AA} + \tfrac{1}{2} z M_B W_{BB} + M_{AB} (W_{AB} - \tfrac{1}{2} W_{AA} - \tfrac{1}{2} W_{BB})$$

$$= U_A + U_B + W M_{AB} . \tag{11.9}$$

Here U_A and U_B respectively are clearly the configurational energy of pure condensed phases of M_A A-molecules and M_B B-molecules and $W = W_{AB} - \tfrac{1}{2} W_{AA} - \tfrac{1}{2} W_{BB}$ is the energy gained on mixing as a result of creating each AB pair of neighbours.

Rushbrooke (1949) shows that the partition function Z for the condensed phase may now be written as a generalisation of (11.7).

$$Z = \sum_{M_{AB}} f_A^{M_A} f_B^{M_B} g(M_A, M_B, M_{AB}) e^{-U_c/kT} \tag{11.10}$$

where $g(M_A, M_B, M_{AB})$ is the number of arrangements of M_A and M_B molecules of each kind such that there are in all M_{AB} pairs of AB neighbours. A solution whose partition function is given by (11.10) is called a **regular** solution.

Using (11.9) Z may be written as

$$Z = (f_A^{M_A} e^{-U_A/kT}) (f_B^{M_B} e^{-U_B/kT}) \sum_{M_{AB}} g(M_A, M_B, M_{AB}) e^{-W M_{AB}/kT}$$

and the crux of the matter is to determine the configurational partition function

$$Z_c = \sum_{M_{AB}} g(M_A, M_B, M_{AB}) e^{-W M_{AB}/kT} . \tag{11.11}$$

The first attempt to treat such a solution, which really means evaluating (11.11), may be called the Bragg-Williams or zeroth approximation. This attempt is equivalent to putting

$$\frac{M_{AA}^* M_{BB}^*}{(M_{AB}^*)^2} = \tfrac{1}{4} \tag{11.12}$$

where M_{AA}^* etc. are the equilibrium values of M_{AA} etc.

We shall not give the full theory of this method but simply state that it predicts positive and negative deviations from Raoult's law for positive and negative values, respectively, of the heat of mixing W. For the case of $W > 0$ there is also a critical temperature below which the solution will separate into two different phases of complementary concentrations x_{1A} and $x_{2A} = 1 - x_{1A}$.

As we have seen in Chapter 11.1 the distinction between a 'poor' and a 'good' solvent is between the two situations in which like pairs of neighbouring molecules are energetically more or energetically less favourable respectively than are unlike pairs of neighbouring molecules. In other words W is respectively positive or negative.

The entropy of mixing ΔS turns out to be the same as that for random mixing considered in Chapter 11.3. In fact the essence of the zeroth approximation, equation (11.12), is that a completely random mixing of the two types of molecules is present in spite of the non-zero value of the heat of mixing.

This assumption of completely random mixing cannot be correct and a better approximation is to write

$$\frac{M_{AA}^* \, M_{BB}^*}{(M_{AB}^*)^2} = \tfrac{1}{4} \, e^{2W/kT} \, . \tag{11.13}$$

This was proposed by Guggenheim and a rigorous derivation was given by Rushbrooke (1938). Since $2W$ is the energy required to change an AA pair and a BB pair into two AB pairs then equation (11.13) is of the type to be expected by analogy with the law of mass action for the reaction

$$AA + BB \rightarrow 2AB \, .$$

For this reason it is often called the **quasi-chemical** approximation. The quasi-chemical approximation leads to partial vapour-pressure curves not dissimilar to those given by the zeroth approximation but the entropy of mixing now obtained is a function of the temperature.

For a fuller discussion of these two approximations the reader is referred to Rushbrooke's *Statistical Mechanics*, Chapter 18 and Guggenheim's *Mixtures*, Chapter 4.

11.5 BINARY MIXTURES OF MOLECULES OF DIFFERENT SIZES

Up to the present we have considered binary mixtures in which the two kinds of molecules are of roughly the same size and shape and so can exchange places with each other. On our quasi-lattice we have in fact regarded each type of molecule as occupying one site.

This quasi-lattice model has been extended to deal with binary mixtures in which one kind of molecule occupies one site and the other kind occupies a number of adjacent sites. A molecule occupying a single site is called a mono-mer. One that occupies two sites is called a dimer, one occupying three sites a trimer, one occupying four sites a tetramer and, in general, one occupying r sites is known as an r-mer. A great deal of work has been done on mixtures consisting of monomers and of larger 'polymer' molecules of the type just described. The main workers in this field have been Chang, Miller, Guggenheim, Rushbrooke, McGlashan, Huggins and Flory.

Much of this work has been done on mixtures which have zero energy of mixing; these are known as **athermal mixtures**. Such mixtures are characterized by the equations $\Delta U = 0$ and $\Delta A = -T\Delta S$. This work is really an extension of the ideal monomer-monomer solution which we considered in Chapter 11.3, and values of such quantities as the free energy of mixing ΔA and the entropy of mixing ΔS have been obtained. Results for monomer-dimer mixtures were obtained by Chang (1939) and for monomer-trimer mixtures by Miller (1942). Mixtures of monomers and of r-mers having the form of an **open** chain (i.e. without any 'closed rings') were studied by Miller (1943), Flory (1942), Huggins (1942) and Guggenheim (1944), who also extended the theory to mixtures of different kinds of open chain r-mers. Guggenheim and McGlashan (1950) have also obtained results for mixtures of monomers with molecules made up of 'closed rings' in the form of triangular trimers and tetrahedral tetramers.

Rushbrooke, Scoins and Wakefield (1953) gave an elegant treatment of athermal monomer-dimer mixtures in which the quasi-lattice model was really pushed to its limit. They derived expressions for the partial vapour pressures of the two components as power series in the volume concentration θ of the dimers. The coefficients of the powers of θ in these expressions are direct analogues of the functions β_k used by the Mayers in the theory of imperfect gases. Trevena (1964) also used this theory for a study of athermal monomer-dimer mixtures and obtained values of the β_k's as far as β_5 for various lattices.

The quasi-lattice approach to a study of athermal mixtures provides a firm basis for a study of mixtures which are not athermal. These latter mixtures are the ones which are met with in practice, and an excellent summary of their properties is given in Chapter 11 of *Mixtures* by Guggenheim. These mixtures are related to athermal mixtures in the same way that regular mixtures are related to ideal mixtures. As for regular mixtures, we can use either the zeroth approximation assuming random arrangement of the molecules, or the first approximation using the equations of quasi-chemical equilibrium. Once again quantities such as ΔA, ΔS can be obtained, and there is a critical temperature below which mixtures split into two phases. Guggenheim discusses the comparison with experiment of some of the results for a mixture of monomers and r-mers. For details the reader is referred to *Mixtures* (p. 232 *et seq*).

In the preceding sections of this Chapter we have discussed the quasi-lattice model in some detail. A great deal of work has been done on it and it is an outstanding example of the success of the methods of statistical thermodynamics. It represents a generalisation to solutions of the lattice or cell model of a pure liquid (see Chapter 4.4).

11.6 EXAMPLES OF DEPARTURES FROM IDEALITY

In the kinetic theory of gases the behaviour of a real gas is usually described by the way in which it differs from that of an ideal gas. In the same way one can consider the departure of the behaviour of real solutions from that of an ideal solution.

First, consider the quantities ΔA, ΔS and ΔU. An example for which the values of these three quantities are very nearly the same as those for an ideal solution is the liquid mixture carbon tetrachloride-cyclohexane. On the other hand, for the mixture acetone-chloroform the values of these three quantities show large deviations from ideality.

Secondly, consider how the vapour pressures of actual liquid pairs depart from the ideal values given by Raoult's law. A few binary miscible liquid mixtures obey Raoult's law for all concentrations, an example being ethylene dibromide — propylene dibromide. On the other hand, a mixture of carbon tetrachloride — cyclohexane shows positive deviations from Raoult's law while a chloroform — acetone mixture shows negative deviations. We have just seen how these deviations can be interpreted as the effects of different sizes of the two molecules and differing interactions between like and unlike molecules.

11.7 MIXTURES OF TWO PARTIALLY MISCIBLE LIQUIDS

When two partially miscible liquids A and B are mixed in various proportions at different temperatures we usually obtain a curve of the type shown in Fig. 11.2.

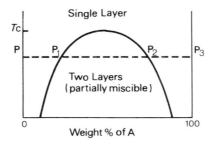

Fig. 11.2 Partial miscibility of two liquids.

To explain the situation, consider what happens at a particular value of the temperature such as that shown by the horizontal dotted line $PP_1P_2P_3$. If we start with the pure B liquid at P and slowly add a small quantity of A, then A

dissolves in B to form a single liquid phase. This happens until we reach the point P_1 when A forms a saturated solution in B. Further addition of A means that the *overall* composition lies between P_1 and P_2, and in practice a second layer starts to form consisting of a saturated solution of B in A of composition P_2. Adding more of A results in an increase in the amount of this second A-rich layer (P_2) and a decrease in that of the B-rich layer (P_1). Eventually, when the *overall* composition reaches that represented by P_2, the B-rich layer disappears, leaving only a single layer of a concentrated solution of B in A. Further addition of A takes us along P_2P_3 where we have a non-concentrated solution of B in A.

As the temperature is raised the same pattern exists, except that points P_1 and P_2 move closer until they merge at a certain **critical solution temperature**. Here the compositions of the two layers become identical, while for higher temperatures the two liquids are completely miscible. We refer to the curve passing through points such as P_1, P_2 for various temperatures as the **solubility-curve**. The region below the curve represents the range in which two mixed liquid layers coexist, and the region above it that of a single mixed layer.

In some cases, for example triethylamine and water, the solubility curve shows a minimum critical temperature instead of a maximum as in Fig. 11.2. In other cases, such as nicotine and water, the solubility curve is a closed curve with an upper and lower critical temperature. Inside the closed curve the two liquids are partially miscible, while outside they are completely miscible. Finally, there are certain mixtures which have neither an upper nor a lower critical solution temperature, an example being ethyl ether and water.

11.8 SOLUTIONS OF GASES IN LIQUIDS

It is probably true to say that a liquid sample without any trace of dissolved gas has never been prepared because all gases are soluble in liquids to some extent. The various factors which affect the solubility, expressed as mass of gas that dissolves in a given mass of liquid, are the nature of the gas, the nature of the liquid, the pressure and the temperature. Gases such as ammonia and hydrogen chloride are very soluble in water. This is because they react chemically with the water to form ammonium hydroxide and hydrochloric acid respectively. Again gases such as oxygen, nitrogen, hydrogen and helium are only slightly soluble in water.

Apart from the very soluble gases the effect of pressure is summed up in **Henry's law**, which states that the solubility of a gas at constant temperature is very nearly proportional to the pressure of the gas above the liquid. Finally, the solubilities of most gases decrease as the temperature rises.

11.9 SOLUTIONS OF SOLIDS IN LIQUIDS

Every solid is soluble to some extent in any liquid even if they do not react chemically with one another, but the solubility is always finite. When the solution is saturated, the limit of solubility has been attained and the saturated solution and excess solid are in equilibrium with one another. There is a wide variation in the concentrations of different solids necessary to produce a saturated solution. This is because the solubilities of two chemically similar substances in the same solvent can be widely different, and the solubilities of the same solute in two chemically similar solvents can be widely different also. For example, we meet a bewildering variety of behaviours if we study the solubilities of the various allotropic modifications of sulphur in various solvents.

In general the solubility of a solid rises rapidly as the temperature increases. We shall see in a moment that this is to be expected, for energy is required to break up the solid and disperse the individual molecules in solution. There are exceptions to this rule, for example anhydrous sodium sulphate becomes less soluble in water as the temperature rises. Since applying pressure has little effect on the density of a liquid, the effect on its structure is also small, and we should expect the effect on solubility of a solid to be small as well. This is observed in practice.

Many of the processes involving solids in practice are carried out in the liquid state. Cooking for example, and other chemical preparations and purfications by leaching and re-crystallization, not to mention metallurgical processes. Obviously knowledge of solubilities of solids is vital to the control of these processes, but much of it remains purely empirical. However, from about the turn of the century some theoretical understanding has gradually been achieved, and we now describe some of the important ideas. One of the authors became interested in solubility curves while studying chemistry at school and the ideas now to be described were beginning to take their present shape at just about this time, the 1930's.

We return to equation (11.11) for the configurational partition function of a regular solution. Assuming for a moment that the molecules of A and B are about the same size, this gives an approximate description of a liquid-liquid mixture; we have

$$Z_c = \sum_{M_{AB}} g(M_A, M_B, M_{AB}) e^{-WM_{AB}/kT} \qquad (11.11)$$

where $W = W_{AB} - \frac{1}{2} W_{AA} - \frac{1}{2} W_{BB}$.

To describe a solid A dissolved in a liquid B, it is only necessary to assume that W_{AA} is negative and numerically larger than the other two energies, in other words that the A molecules 'like being together'. Suppose also for definiteness that $M_A < M_B$. Then M_{AB}, the number of unlike neighbour pairs in equation

(11.11), may take any value from effectively zero (which corresponds to the formation of separate A and B phases) to zM_A (which means that each A molecule is completely surrounded by B molecules). To evaluate (11.11) rigorously would need a knowledge of the function g and how it varies with M_{AB}. Precise information is available only in certain special cases, for example if $M_A = M_B$, and we have to resort to approximations.

Consider first of all the case of negative W, that is to say where unlike atoms prefer to be together. The free energy of mixing is obtained from (11.11) by taking $-kT \, \ln(Z_c)$, which is approximately equal to $-kT \, \ln g^* + WM_{AB}^*$. Here M_{AB}^* corresponds to the maximum term in (11.11), g^* being the value of g for this value of M_{AB}. By the rules of statistical mechanics, the value of M_{AB} corresponding to the **maximum term** in (11.11) is the **equilibrium** value of this quantity, which we call M_{AB}^*. If W is negative we would expect M_{AB}^* to be at or near its maximum possible value zM_A. The assembly will consist of a random uniform mixture of the two types of molecule. Substances A and B will mix in all proportions, and chemists would say that they are 'good' solvents of one another.

To get finite mutual solubility we must have W positive, and we have pointed out above that this is to be expected if A is a solid and B a liquid. If the assembly is of uniform composition everywhere, the value M_{AB}^* that gives the maximum term in (11.11) will be somewhere near the middle of its possible range. In thermodynamic terms WM_{AB}^* is the energy of mixing and $k \, \ln g^*$ is the entropy of mixing. If we increase M_{AB}^* we gain entropy but have to supply some energy, in other words there is a conflict and we must have M_{AB}^* less than zM_A, that is to say that the mixing is not complete. As we vary the temperature, the relative importance of the energy and entropy terms in the free energy obviously changes.

At very high temperatures the entropy term dominates and we again have nearly random mixing. At very low temperatures the energy term dominates and the most likely configuration is a two-phase one in which all the A molecules and all the B molecules are together. Thus at a low enough temperature we might expect to be able to gain free energy by splitting the assembly into two different phases, one of them rich in A, the other poor in A. Using our approximate knowledge of the behaviour of the function g we can work out the mathematical details and arrive at conclusions very similar to those summarised above (Chapter 11.6) for partially miscible liquids. However, we expect the solubility curve to be decidedly more asymmetric for a solid and a liquid than it is for two liquids. In addition, we do not expect a consolute temperature unless solid A melts before liquid B boils, in which case we should be back to the situation of two partially miscible liquids. We refer the reader to textbooks on statistical mechanics for the mathematical details, which are fairly involved even if we make the simplest possible realistic approximations for $g(M_A, M_B, M_{AB})$.

This work does enable us to gain insight into one puzzling fact, the widely different solubilities of the different modifications of sulphur. We have seen above that estimates can still be made of the behaviour of the function g even when we have molecules of different sizes. In particular, various workers have studied the effect of assuming that one of the types of molecule can aggregate into dimers, chains or rings. Without going into details, one can easily see that the function g *must* be decreased by such constraints. Obviously two molecules forming a dimer can be arranged in the assembly in a much smaller number of ways than can two molecules that wander around independently. Decreasing g is going to have the same effect on the free-energy balance as increasing W, and we know that increasing the latter must reduce mutual solubility. We can now understand qualitatively the differing solubilities of the different modifications of sulphur, since the differences between them are believed to be explained by the formation of 'ring molecules' and polymer-like chains of sulphur atoms. We can also understand qualitatively why the solubilities of molecules like the paraffins usually decrease as the chains become longer. Further progress will be possible as we learn more about the function $g(M_A, M_B, M_{AB})$ for various models.

11.10 COLLIGATIVE PROPERTIES OF SOLUTIONS

We shall now consider solutions consisting of a solvent and nonvolatile solute. In this case the total vapour pressure of the solution is the vapour pressure of the solvent. Such a solution has certain properties which depend on both the nature of the solvent and the concentration of the solution but do not depend in any other way on the properties of the solute; these are called the **colligative** properties. The name wrongly suggests that these properties are related to 'the tendency of the solute molecules to stick together.' It cannot be stressed too strongly that these properties would occur even in an ideal solution and that the effect on them of the relative strengths of interaction of solute and solvent molecules is small.

The colligative properties of nonelectrolytic solutions are important because they provide methods for estimating the molecular weight of the dissolved solute. In recent years solutions of polymers and proteins have been studied and information can now be obtained about the average number of atoms in a polymer molecule, known as the 'length of an average chain' or its 'effective molecular weight'. Such information is often crucial in polymer physics or molecular biology.

The four colligative properties will now be discussed briefly.

11.10.1 The lowering of the vapour pressure

From Raoult's law (equation (11.4)) the vapour pressure of an ideal dilute solution is given by

$$P = x_A P^0$$

where x_A is the mole fraction of the solvent. Furthermore, as the solute is non-volatile, P and P^0 represent the vapour pressure of the solution or the solvent. Since $x_A + x_B = 1$, where x_B is the mole fraction of the solute

$$P = (1 - x_B)P^0$$

or
$$\frac{P^0 - P}{P^0} = x_B .$$ (11.14)

Thus we see that the presence of the solute causes a relative decrease in the vapour pressure of amount $(P^0 - P)/P^0$, which depends only on the concentration of the solute. This relative lowering of vapour pressure is therefore a colligative property.

11.10.2 The elevation of the boiling point

Solutions which contain nonvolatile solutes boil at temperatures higher than the boiling point of the pure solvent. This can be explained in terms of the vapour pressure lowering since it is a direct consequence of it. The boiling point is the temperature at which the vapour pressure is equal to the ambient pressure, usually atmospheric. We have just seen that the addition of a nonvolatile solute lowers the vapour pressure, so a higher temperature has to be attained before the vapour pressure of the solution becomes equal to the ambient pressure. Vapour pressure changes rapidly with temperature, but for a dilute solution the temperature change involved is small, and we can write $\Delta T = \Delta P \times \dfrac{dT}{dP}$ where dT/dP is the slope of the vapour pressure curve. This is connected with measurable properties of the liquid by the Clausius-Clapeyron equation (equation (5.2)). Thus the added solute produces an elevation of the boiling point which can be shown to be independent, at least in dilute solutions, of all properties of the solute except its mole fraction in the solution. It is thus a colligative property.

11.10.3 The lowering of the freezing point

When a dilute solution is cooled a temperature will be reached eventually at which **solid solvent** is formed. This temperature, at which such a separation begins, is known as the freezing point of the solution. At this temperature the solution and solid solvent coexist.

Solutions freeze at temperatures *lower* than the freezing point of the pure solvent, and this lowering of the freezing point is again a consequence of the vapour pressure lowering. To see this we refer to Fig. 11.3. In this diagram,

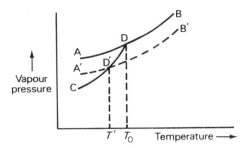

Fig. 11.3 Lowering of freezing point by a solute.

AB is the vapour pressure curve of the pure liquid solvent and CD the sublimation curve of the solid solvent. At the freezing point of the pure solvent the solid and liquid phases are in equilibrium and therefore have the same vapour pressure since either of them is in equilibrium with the same vapour phase. The only point on the two curves at which the vapour pressures of both solid and liquid are equal is the point D where the curves AB and CD intersect. Thus the temperature T_0 corresponding to D is the freezing point of the pure solvent. When we have a solute present the vapour pressure is lowered and the vapour pressure curve of the solution is a curve such as A'B' lying below AB. This curve A'B' cuts the sublimation curve at the point D', where the corresponding temperature T' is now the new, but lower, freezing point. For small $(T_0 - T')$ the relevant portions of the curves AD, A'D', D'D are nearly straight, and we conclude that the lowering of freezing point is proportional to the lowering of vapour pressure in a dilute solution.

The freezing point depression $(T_0 - T')$ is shown in books on physical chemistry to be independent of any property of the solute except its concentration in the solutions. It is therefore another colligative property.

11.10.4 Osmosis

If a solution and solvent are separated from each other by a semipermeable membrane that will permit the passage of solvent but not of solute molecules, the solvent molecules will try to find their way into the solution. If the solution is contained in a closed vessel a very considerable pressure may then be built up in it. The basic principle is shown in Fig. 11.4. This persistent tendency of a solution to dilute itself is known as **osmosis**. The pressure which would have to be applied to the solution to *prevent* any osmosis of the pure solvent into it is called the **osmotic pressure**, Π, of the solution. It must again be emphasised that this is a property of both ideal and non-ideal solutions.

There is a great variety of semipermeable membranes. Cellophane and certain protein membranes, for example, do not allow the passage of high molecular weight solutes but are permeable to water. In the early days it was found that a

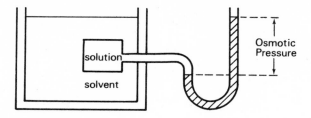

Fig. 11.4 The principle of osmotic pressure.

deposit of copper ferrocyanide in the pores of a porous pot (called a Pfeffer pot) acted as a semipermeable membrane for low molecular weight solutes in water. Again palladium metal foil is permeable to hydrogen gas but not to nitrogen gas.

Suppose we introduce sugar solution into a closed Pfeffer pot and have plain water surrounding the pot in an arrangement like that in Fig. 11.4. What is observed in practice is that the water diffuses into the pot and a large pressure builds up inside it. It is natural to think of this as a consequence of an attraction between the water and sugar molecules but we have pointed out already that this is misleading because osmotic pressure would occur with an ideal solution. It is really better to think of the excess pressure as being exerted by the sugar molecules, which have access to only one side of the wall of the pot and are turned back whenever they approach it. By analogy with the kinetic theory of gases, whenever a solute molecule collides with the walls of the pot and is turned back there must be a force and pressure associated with this continual transformation of momentum.

Osmotic pressure may vary from a few millimetres of mercury for a dilute solution to many hundreds of atmospheres for a concentrated solution of a very soluble substance such as sugar. A direct measurement of osmotic pressure is not easy and the reader is referred to textbooks on physical chemistry for the details. The usual way of interpreting the osmotic pressure of a solution is to relate it to the vapour pressure lowering and the other colligative properties. We shall now do this from purely thermodynamic considerations.

Suppose then that we have solution and pure solvent separated by a membrane. The molar free energy of the solvent in the solution is less than the molar free energy of the pure solvent on its own. Thus the solvent tends to move from the high free energy state of the pure solvent to the lower free energy state of the solution. This tendency can be resisted if the free energy of the solution is increased by applying an excess pressure Π to the solution. Books on physical chemistry show that the free energy *decrease* per mole of solvent that occurs when a solute is added is $RT \ln(P/P^0)$ where P^0 and P are the vapour pressures of the pure solvent and solution respectively. Furthermore the free energy *increase* per mole of solvent due to the excess pressure Π is Πv where v is the volume of 1 mole of solvent in the solution. Thus for equilibrium

$$\Pi v + RT \ln \frac{P}{P^0} = 0$$

or
$$\Pi v = - RT \ln \frac{P}{P^0} . \qquad (11.15)$$

This equation relates the osmotic pressure, Π, to the lowering of the vapour pressure of a solution. It is important to realize that the above argument makes no assumption at all about the mechanism causing osmotic pressure or the nature of the membrane. It is a purely thermodynamic argument.

For a dilute solution the solvent obeys Raoult's law, that is $P/P^0 = x_A$, and so on substituting in (11.15) we obtain

$$\Pi v = - RT \ln x_A . \qquad (11.16)$$

This shows that the osmotic pressure Π depends on the temperature T, the concentration and the molar volume v of the solvent in the solution, but not on any properties of the solute. Therefore it is a colligative property. Note also that for such dilute solutions, the volume of one mole of solvent in the solution is essentially equal to the molar volume of the pure solvent.

Since $x_A + x_B = 1$, then

$$\ln x_A = \ln (1 - x_B)$$

$$\hateq - x_B$$

if $x_B \ll 1$, that is for a dilute solution. So using (11.16) we have

$$\Pi v = RT x_B . \qquad (11.17)$$

If there are respectively n_A and n_B moles of solvent and solute in a given quantity of solution, then

$$x_B = \frac{n_B}{n_A + n_B}$$

$$\hateq \frac{n_B}{n_A} \quad \text{for a dilute solution.}$$

Thus (11.17) becomes

$$\Pi n_A v = n_B RT$$

or $$\Pi V = n_B R T \qquad (11.18)$$

where $V = v n_A =$ the volume of solvent containing n_B moles of B

$\hateq$ the volume of solution containing n_B moles of B.

Equation (11.18) is known as **van't Hoff's law** for ideal solutions. It has a form similar to the ideal gas law with Π replacing the gas pressure P, and can be used to find molecular weights of dissolved solutes just as the ideal gas law can be used to find the molecular weights of gases. However, because of the difficulty in making accurate osmotic pressure measurements for dilute solutions, this method is usually employed only for solutes consisting of long polymer molecules. In most cases the molecular weights of solutes are obtained from a knowledge of one of the other colligative properties.

Equation (11.18) can also be deduced by the following argument. Osmotic pressure is, as we have seen, the excess pressure exerted by the solute molecules. In a dilute solution the solute molecules are far apart and move almost independently of one another, and each of them contributes to the pressure as it reaches the boundary of the vessel and is turned back. It is therefore reasonable that the law relating pressure to concentration should be similar to the relation between pressure and density in a perfect gas. However the thermodynamic argument has the advantage of showing that the constants are the same.

11.11 ELECTROLYTIC SOLUTIONS

Although apparently more complicated than non-electrolytic solutions, electrolytes are in fact much better understood. The first step was taken by Debye and Hückel, who combined electrostatics and statistical mechanics to find an approximate description of the 'atmosphere' of oppositely charged ions surrounding an ion in the electrolyte. This work was improved by Onsager, Kirkwood and many other workers, and both equilibrium properties and electric conductivity can usually be predicted with good accuracy. In this context, the properties of the ionic atmosphere are more important than those of the liquid distribution function. Thus it is not usually necessary to consider the latter explicitly, the thickness of the atmosphere usually being large compared with ionic diameters. For an account of this field the reader is referred to Harned and Owen *The Physical Chemistry of Electrolytic Solutions* (Reinhold, New York, 1950).

REFERENCES

Books
Barrow, G. M., 1966, *Physical Chemistry, 2nd Edition,* McGraw-Hill Book Company.

Denbigh, K., 1970, *Principles of Chemical Equilibrium, 2nd Edition,* Cambridge University Press.

Guggenheim, E. A., 1952, *Mixtures,* Oxford University Press.

Harned, G. S. and Owen, B. B., 1950, *The Physical Chemistry of Electrolytic Solutions,* Reinhold, New York.

Hildebrand, J. H. and Scott, J. L., *Solubility of Non-electrolytes, 3rd Edition,* Dover, New York.

Maron, S. H. and Prutton, C. F., 1965, *Principles of Physical Chemistry, 4th Edition,* The Macmillan Company, New York.

Miller, A. R., 1948, *The Theory of Solutions of High Polymers,* Oxford University Press.

Moelwyn-Hughes, E. A., 1961, *Physical Chemistry, 2nd Revised Edition,* Oxford.

Rowlinson, J. S., 1969, *Liquids and Liquid Mixtures, 2nd Edition,* Butterworths.

Rushbrooke, G. S., 1949, *Introduction to Statistical Mechanics,* Oxford University Press.

Temperley, H. N. V., 1963, *Properties of Matter, 2nd Edition,* University Tutorial Press, London.

Papers

Chang, T. S., 1939, *Proc. Camb. Phil. Soc.,* **35**, 265.

Flory, P. J., 1942, *J. Chem. Phys.,* **10**, 51.

Guggenheim, E. A., 1944, *Proc. Roy. Soc.,* **A183**, 203.

Guggenheim, E. A., 1944, *Proc. Roy. Soc.,* **A183**, 213.

Guggenheim, E. A. and McGlashan, M. L., 1950, *Proc. Roy. Soc.,* **A203**, 435.

Huggins, M. L., 1942, *Ann. New York Acad. Sci.,* **43**, 9.

Miller, A. R., 1942, *Proc. Camb. Phil. Soc.,* **38**, 109.

Miller, A. R., 1943, *Proc. Camb. Phil. Soc.,* **39**, 54.

Rushbrooke, G. S., 1938, *Proc. Roy. Soc.,* **A166**, 296.

Rushbrooke, G. S., Scoins, H. I. and Wakefield, A. J., 1953, *Disc. Faraday Soc.,* No. **15**, 57.

Trevena, D. H., 1964, *Proc. Phys. Soc.,* **84**, 969.

Chapter 12
TRANSPORT PROCESSES

12.1 INTRODUCTION

As the name implies a transport process in a medium occurs when something is transferred from one part of the medium to another. The most familiar transport processes are viscosity and thermal conduction, and they involve the transport of momentum and energy, respectively, through a medium. For gases it is possible, on the basis of the kinetic theory, to derive expressions for both the viscosity and thermal conductivity in terms of molecular properties. The reader is referred to text-books on properties of matter for the details. As we shall see, a corresponding derivation is not such an easy matter for a liquid. It is clear, however, that transport processes are *non-equilibrium* processes in either a gas or liquid. Moreover we should remember that up to the present we have been considering liquids in thermodynamic *equilibrium*. In equilibrium a given property is assumed to be the same throughout the whole volume of the liquid and also to be constant with time except for small fluctuations. Consider however what happens when a property — say temperature — is *not* uniform throughout a liquid at a given time. When this occurs heat energy is transported from the hotter to the colder regions, and the resulting thermal conduction is a transport process in which energy flows in a direction opposite to that of the temperature gradient. Again consider viscosity, which comes into play when adjacent parts of the liquid move with different velocities, in other words, when a velocity gradient is present. The viscous forces involved act so as to cause the slower moving regions to move more rapidly and the faster moving ones more slowly. This tendency to 'even out' the velocities involves a transport of momentum in a direction perpendicular to the fluid velocity (see Chapter 6.8). There is a third well-known transport process, namely ordinary diffusion, which is the transport of mass from one region to another because of a gradient in the density. In this Chapter we shall be mainly concerned with viscosity and, to a lesser extent, heat conduction.

12.2   MICROSCOPIC REVERSIBILITY AND
MACROSCOPIC IRREVERSIBILITY

In the kinetic theory of transport processes the motion of the molecules is completely reversible; we can go back as well as forward in time because of the symmetry of the laws of dynamics as applied to the individual molecules. In other words these laws and the equations of motion apply just as well to the molecular motions *in reverse*. Because of these considerations we have what can be called **microscopic reversibility**. On the other hand, at the macroscopic level transport properties are not reversible. For example, in a liquid in which a temperature gradient exists the temperature will, in time, become uniform throughout the liquid due to the conduction of heat from the hotter to the cooler regions. However the converse will not happen: a liquid which is at a uniform temperature throughout will not spontaneously split up into finite regions of different temperatures. Again, a liquid which is initially stirred so as to produce in it a rotary motion will settle down in time. On the other hand a liquid in equilibrium will not, conversely, suddenly start to swirl around. These two are examples of **macroscopic irreversibility**.

So we have a conflict, a paradox — microscopic reversibility on the one hand and macroscopic irreversibility on the other. The motion of the molecules is reversible, but the transport processes observed macroscopically are clearly irreversible. We therefore need to find a 'bridge' or link to take us from the **reversible** equations of molecular dynamics to the observed macroscopic **irreversible** trend towards equilibrium, that is, irreversibility must somehow be introduced into the microscopic picture. This bridge is what we need to resolve the paradox.

In a gas the molecules spend most of their time travelling through empty space between collisions. We then need an assumption which introduces the required irreversibility. One form of this assumption is the molecular chaos hypothesis, which states that the momentum of a typical molecule at a time t_2 is statistically independent of its momentum at an earlier time t_1, even though only a few collisions occur between t_1 and t_2. For liquids the individual molecules do not travel around in largely empty space, as do the gas molecules, but are confined to 'cells' formed by ten or so nearest neighbours. So another assumption is required, one which will be a counterpart of the molecular chaos hypothesis. Kirkwood (1946) proposed such an assumption, and we shall return to this later as it is the 'bridge' that we require.

In passing, we might note that reversible and irreversible processes are also considered in thermodynamics, where the irreversible evolution towards equilibrium is embodied in the principle of the increase of entropy. This tells us that an isolated assembly tends to a state of maximum entropy at equilibrium.

12.3   TRANSFER OF MOMENTUM AND ENERGY
IN GASES AND LIQUIDS

Let us now look a little more closely at the mechanisms involved in transport processes in gases and liquids. In a gas a molecule spends most of its time in empty space between collisions; hence the transport of momentum and energy is accomplished by each molecule carrying its own momentum and energy with it as it thus moves. Let us call this mechanism (1). In a liquid things are quite different. Each molecule is enclosed in a cell formed by its nearest neighbours, and only occasionally migrates to a neighbouring cell. It spends most of its time in this cell and is therefore under the influence of its intermolecular interactions with its neighbours. The effect of these interactions is to give rise to 'drag' which causes the fast molecules to be slowed down while the slower ones are speeded up. Consequently the momentum and energy are transferred *through space* from the faster to the slower molecules. Let this be referred to as mechanism (2).

Let us now concentrate on viscosity and confine our discussion to laminar flow in a simple Newtonian liquid. Here molecules interact by means of central forces only, that is, spherically symmetrical forces of the type we considered in Chapter 3.2. Such laminar flow was considered in Chapter 6.8 and is associated with two mutually perpendicular directions. Rewriting equation (6.53), we have

$$P_{xy} = \eta \, \frac{\partial v_x}{\partial y} \, , \qquad (12.1)$$

Here η is the coefficient of shear viscosity. P_{xy} is the shear stress in the x-direction perpendicular to the y-axis. The first suffix x indicates the direction of this stress and the second one y that of the axis perpendicular to the plane in which it acts. The quantity $\partial v_x/\partial y$ is the velocity gradient showing how the x-component of velocity varies in the y-direction.

We must now discuss how to obtain an expression for the transfer of x-momentum in the y-direction using the microscopic approach based on the molecular mechanisms (1) and (2) mentioned earlier. To do this we return to consider Fig. 6.8. As the top plate moves to the right it imparts momentum to the liquid in contact with it. Since force is rate of change of momentum, the momentum given to this top layer of liquid in time δt is $F\delta t$, where F is the force acting on the upper plate. The lower fixed plate will experience an equal and opposite force F because of the dragging force due to the liquid; thus the liquid in contact with this lower plate will be deprived of a momentum of $F\delta t$ in this time. Hence the resultant effect is as though a momentum of $F\delta t$ is transferred a distance a through the liquid in the y-direction between the plates. This rate of transfer of momentum is (the magnitude of momentum transferred) × (velocity of transfer),

that is
$$F\delta t \times \frac{a}{\delta t}$$

$$= Fa$$

$$= P_{xy}\, Aa$$

$$= P_{xy}\, V$$

where V is the volume of liquid between the plates each of area A. If V happened to be *unit* volume, this rate of transfer of x-momentum in the y-direction would be simply P_{xy}, numerically equal to the shear stress. Let us consider unit volume of a liquid in which N molecules occupy a volume V.

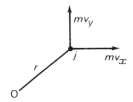

Fig. 12.1 Momentum transfer arising from the mechanism (1) in which each molecule carries its own momentum with it as it moves.

Mechanism (1)

Referring to Fig. 12.1 we consider a molecule j placed at the point (x, y) at a distance r from the origin. This molecule has momentum components of $p_x = mv_x$ and $p_y = mv_y$ as shown. It carries its own x-momentum in the y-direction at the rate v_y, thus giving a contribution of $p_x v_y = p_x\, p_y/m$ to this momentum transfer. The total contribution to the macroscopic momentum transfer due to mechanism (1) is then obtained by summing over all the molecules in unit volume.

Mechanism (2)

As already mentioned, this mechanism arises from the presence of the intermolecular interactions. We shall first consider only two molecules i and j at distance r apart, the former being at the origin and the latter at the point (x, y) as in Fig. 12.2. In the notation of Chapter 3 the force between these molecules along the line joining them is $F(r) = -d\phi/dr$. The x-component of this force is

$$F_x = F(r)\cos\alpha \tag{12.2}$$

and is the rate of change of x-momentum of the molecule j. (α is the angle between the x-axis and the line joining the two molecules). Thus in a time δt *the molecule j* will suffer a change of amount $F_x \, \delta t$ in its x-momentum due to the interaction of molecule i at the origin. Similarly, by Newton's third law this

Fig. 12.2 Momentum transfer arising from the interaction of two molecules (mechanism 2).

molecule at the origin will suffer an equal but opposite change in x-momentum due to the molecule j at (x, y). Thus we can regard this mechanism as being equivalent to a transfer of x-momentum of $F_x \, \delta t$ through space from molecule i to molecule j in time δt, that is through a distance y in the y-direction in time δt. Hence the rate of transfer of x-momentum in the y-direction is

$$F_x \, \delta t \times \frac{y}{\delta t} = F_x y$$

$$= F(r) \cos\alpha \times y$$

$$= F(r) \frac{xy}{r} \, . \tag{12.3}$$

The total contribution to the macroscopic momentum flow due to mechanism (2) is obtained by summing expression (12.3) over each pair of molecules in unit volume.

Hence, if we take the total macroscopic momentum flow P_{xy} due to both mechanisms (1) and (2) we can write

$$P_{xy} = \sum_{(1)} \frac{p_x p_y}{m} + \sum_{(2)} F(r) \frac{xy}{r} \tag{12.4}$$

where the first summation is over all the N/V molecules in unit volume and the second over all $N^2/2V^2$ pairs of molecules in unit volume.

Equation (12.4) includes both mechanisms (1) and (2) and is applicable to both gases and liquids. Since, as we have seen, mechanism (1) is the one mainly responsible for the transport of momentum in gases, the most important contribution to P_{xy} in gases comes from the first summation $\sum_{(1)}$ on the right-hand side of this equation. For its evaluation a knowledge of f_N, the complete distribution function for the co-ordinates and momenta of all the N molecules, is necessary. In liquids the first term can be neglected and the overwhelming contribution comes from the second summation $\underset{(2)}{\Sigma}$, whose evaluation depends on summing over pairs of molecules. As we shall see, this means knowing the radial distribution function under non-equilibrium, that is steady flow conditions, and we shall discuss this later.

Above we have given an outline of the method of obtaining the macroscopic momentum transfer in terms of the microscopic, or molecular, properties. This momentum transfer is the one involved in the viscosity. We have considered the particular case of laminar flow only, but results giving the momentum transfer P for more general types of liquid flow have been obtained as well, in addition to general expressions for the flow of energy Q involved in heat conduction. Both P and Q vanish for liquids in equilibrium, but for the non-equilibrium conditions accompanying velocity or temperature gradients P and Q are non-zero, and we need to use non-equilibrium distribution functions for their evaluation. Our next step must therefore be to discuss how this may be done.

12.4   NON-EQUILIBRIUM DISTRIBUTION FUNCTIONS

In the theory of liquids in equilibrium the various distribution functions form a hierarchy starting with f_N, the distribution function describing the co-ordinates and momenta of all the N molecules in our liquid of volume V. If we average over all but two of the molecules we obtain n_2, the pair distribution function. Introducing an appropriate normalising factor we can derive the radial distribution function $g(r)$ from n_2. This hierarchy can be represented thus:

$$f_N \rightarrow n_2 \rightarrow g(r).$$

We now consider how these distribution functions must be modified in non-equilibrium conditions. For this purpose we will concentrate in particular on the number density $\rho(r) = \rho_0\, g(r)$ (see Chapter 2.3) and discuss how it is changed when a laminar flow exists in the liquid. The function $\rho(r)$, which is spherically symmetrical in equilibrium conditions, must now be replaced by some function which depends not only on r but on direction as well. Using any chosen representative molecule as the centre of a spherical polar coordinate system, $\rho(r)$ is then replaced by $\rho(r, \theta, \phi)$. The number of molecules in a small volume dV at (r, θ, ϕ) will now be $\rho(r, \theta, \phi)dV$, not $\rho(r)dV$ as in equilibrium.

To obtain the momentum flow P_{xy} from equation (12.4) we need to find the sum $\underset{(2)}{\Sigma}$ over all the pairs. We therefore concentrate on our typical molecule i at the origin and consider the pairs which it forms with each molecule in our small volume dV. The number of such pairs is equal to the number of molecules in this volume, namely $\rho(r, \theta, \phi)dV$. Furthermore all the molecules in this small volume have virtually the same coordinates (x, y, z) or (r, θ, ϕ) with respect to the origin, and hence the contribution to the sum $\underset{(2)}{\Sigma}$ from all these pairs is

$$\rho(r, \theta, \phi)\mathrm{d}V \ F(r) \ \frac{xy}{r} \ .$$

Remembering that $\quad x \ = \ r \sin\theta \ \cos\phi$

$$y \ = \ r \sin\theta \ \sin\phi$$

and $\qquad\qquad\quad \mathrm{d}V \ = \ r^2 \sin\theta \ \mathrm{d}\theta\mathrm{d}\phi\mathrm{d}r$

this contribution can be written as

$$\rho(r, \theta, \phi) \, F(r) \, r^3 \, \sin^3\theta \ \sin\phi \ \cos\phi \ \mathrm{d}\theta\mathrm{d}\phi\mathrm{d}r \, .$$

To obtain P_{xy} this expression is integrated over the volume V of the liquid and this is then repeated for each of our N molecules selected as the origin molecule in its turn. After this we divide by two because each ij molecular pair has been counted twice and finally we divide by V since P_{xy} refers to unit volume. The integral expression for P_{xy} thus obtained involves $F(r), \rho(r, \theta, \phi)$ and the various spherical polar coordinates. To simplify matters various workers have assumed that for laminar flow with a small velocity gradient $G_{xy} = \partial v_x/\partial y$, the non-equilibrium number-density distribution function can be written as

$$\rho(r, \theta, \phi) \ = \ \rho(r)\,[1 \, + \, G_{xy}\,u(r)\sin^2\theta \ \sin\phi \ \cos\phi] \ .$$

Here $u(r)$ is a function we shall discuss below and the second term in the square bracket is small compared with unity. With this amendment the expression obtained for P_{xy} involves $F(r)$, $\rho(r)$ and $u(r)$. Using equation (12.1) we can finally obtain an expression for η. This turns out to be

$$\eta \ = \ \frac{2\pi}{15} \, \frac{N}{V} \int \frac{\mathrm{d}\phi(r)}{\mathrm{d}r} \ \rho(r)\,u(r)\,r^3 \ \mathrm{d}r \qquad\qquad (12.5)$$

where $F(r)$ has been written instead as $- \, \mathrm{d}\phi(r)/\mathrm{d}r$.

Let us now consider this expression for η. It contains three functions of r, namely $\phi(r)$, $\rho(r)$ and $u(r)$, which we will discuss in turn.

(a) $\phi(r)$, the intermolecular potential function can, for this purpose, be represented by the LJD form discussed in Chapter 3.

(b) $\rho(r)$, the equilibrium distribution, was discussed in Chapter 2.3 and it is reasonably well known from work on X-ray and neutron diffraction.

(c) We are left with $u(r)$, about which we know little except that it represents the radial distortion caused by the non-equilibrium conditions produced by the velocity gradient. Let us bear in mind however that non-equilibrium conditions can be produced by temperature and density gradients also, in thermal conduction and diffusion respectively. In other words the function $u(r)$ is of the type involved in the more general problem of the amended distribution functions appertaining to non-equilibrium conditions — whether such a non-equilibrium state is produced by a gradient of velocity, temperature or density. It is also tied up with the problem of statistical averaging to which we now turn.

12.5 AVERAGING PROCESSES AND KIRKWOOD'S HYPOTHESIS FOR LIQUIDS

In the last section we considered the equilibrium distribution function $\rho(r)$ and noted that it could be derived from the equilibrium distribution function f_N for all the coordinates and momenta of a liquid of N molecules. Under non-equilibrium conditions, where events vary with time, the function f_N will also vary with time; to emphasize this fact we write it as $f_N(t)$. This function $f_N(t)$ satisfies a certain formal differential equation in the coordinates, the momenta and the time known as the Liouville equation. This equation essentially describes the behaviour of the assembly of N molecules according to Newton's laws of motion and it is **reversible**, allowing for molecular motions in both forward and reversed directions. In other words, this equation describes the microscopic reversibility which we discussed earlier in Chapter 12.2.

Detailed study of the properties of this equation by Prigogine and his co-workers have led to a solution of the 'irreversibility paradox'. Any *finite* assembly will return very closely to its initial configuration after a definite time, the so-called Poincaré period. This period is, however, exceedingly long. For a real assembly containing say 10^{20} molecules, the 'time of doing an experiment' may be of the order of a fraction of a second up to days. This is always extremely short compared with the Poincaré period of the order of millions of years. To get *truly* irreversible behaviour we need an infinite assembly but, for the above reason, the behaviour of a real assembly is not measurably different. There is little difference in practical terms between these two statements:

(a) In a real assembly, a spontaneous decrease of entropy never occurs.

(b) In a real assembly one would have to wait an extremely long time for a fluctuation corresponding to a decrease of entropy of appreciable magnitude.

The work of Prigogine and his school confirms the hypothesis made by Kirkwood which will be explained below. In effect, we can avoid the difficulty of principle if we are careful about how we average over time.

To go from a molecular to a macroscopic description of the liquid involves a statistical averaging process. Perhaps the most familiar example, in elementary physics, of the use of such an averaging process is in the kinetic theory of gases. In this theory we start from the microscopic standpoint and consider the motions of the individual molecules in a cubical box. Then we introduce certain averaging procedures such as a mean square velocity and the assumption that on average, only one-sixth of the molecules will be moving normally towards one wall of the box, and so on. After such averaging procedures we end up with macroscopic equations such as $PV = RT$, etc. So such averaging processes imply some concealed assumption or hypothesis which only allows the irreversible evolution to equilibrium observed on the macroscopic level in spite of the un-doubted reversibility on the microscopic level.

In gas theory the assumption which introduced the required irreversibility was the molecular chaos hypothesis (see Chapter 12.2). The success of this theory for gases supports the idea that irreversibility is connected with the breakdown of statistical correlation between the successive values of the magni-tude and direction of the momentum of a given molecule separated by a certain interval of time.

As already mentioned in Chapter 12.2, Kirkwood introduced a corres-ponding hypothesis which is the basis of his work on irreversible processes in liquids. It is concerned with the resultant force $\mathbf{F}(t)$ which acts on a given representative molecule of the liquid. He then considers how to describe mathe-matically the statistical correlation between the values $\mathbf{F}(t)$ and $\mathbf{F}(t + t')$ of this force at times t and $t + t'$. To do this he assumes that at a time $t + \tau$ the force on the molecule will be statistically independent of its earlier value at time t. In other words, at the time $t + \tau$ the molecule has 'forgotten' the force that acted on it at the beginning of the interval at time t. This ability of the molecule to 'remember' or 'forget' what happened to it at an earlier time τ is measured by the autocorrelation function of $\mathbf{F}(t)$ defined by

$$\xi \;=\; \overline{\mathbf{F}(t).\mathbf{F}(t + t')} \qquad\qquad (12.6)$$

To obtain ξ we form the scalar product of the vectors $\mathbf{F}(t)$ and $\mathbf{F}(t + t')$ and calculate the average value of this product for all the molecules, this ensemble average being denoted by a bar. Kirkwood's hypothesis then states that:
(1) If $t' < \tau$, $\xi \neq 0$, that is, the force on the molecule is not statistically independent of its previous value $\mathbf{F}(t)$ at time t. We could say that the molecule 'remembers' the force which acted on it at this earlier time.
(2) If $t' > \tau$, $\xi = 0$, that is, the force on the molecule is statistically inde-pendent of its value $\mathbf{F}(t)$ at time t. The molecule can now be said to have 'forgotten' the force which acted on it at this earlier time.

The actual force $\mathbf{F}(t)$ acting on a given molecule is the resultant of all the forces on it due to all the other molecules. In practice, of course, this force is mainly due to the pairs it forms with its nearest neighbours, with a much smaller contribution coming from pairs formed with the remaining molecules. $F(t)$ fluctuates with time as all these molecules move around in their molecular motions.

To sum up, the essence of Kirkwood's hypothesis can be stated thus. The force $F(t)$ on one molecule slowly loses correlation with its original value in such a way that after a time τ has elapsed, the autocorrelation function $F(t).F(t + \tau)$ has decayed effectively to zero. In addition it is assumed that τ is small compared with macroscopic times, the time for which one observes the macroscopic behaviour of the liquid in any actual experiment.

We have already mentioned that the resultant force $F(t)$ acting on our representative molecule will vary with time because $F(t)$ depends on the relative positions of the molecules surrounding this molecule and these positions vary continuously due to the motion of these molecules. This applies to our representative molecule also because it, too, is moving like all the others. As it so moves around, its very motion brings into operation a 'frictional' force which on the whole tends to oppose its motion. These phenomena have been compared with the Brownian motion of a solid particle suspended in a liquid. Such a particle moves continually in all directions, and its resultant motion is pictured as being due to two opposing influences. First, the particle is caused to move by the bombardment of its surface by the surrounding liquid molecules. During a small period of time there will be more impacts on one side than on the other and the particle will therefore move in the direction of the excess force until its path is later changed by subsequent molecular impacts. In this way the particle moves in a random zig-zag path. Secondly, its motion is opposed by a frictional force which can be described as follows. The size of the suspended particle is much greater than that of the surrounding liquid molecules. Thus the particle can be regarded as being in an environment which is a continuous medium and a 'hydrodynamic' picture is valid. These considerations lead to Stokes' law, which tells us that the opposing force on a spherical particle of radius a is $6\pi\eta av$, where η is the viscosity of the liquid and v the velocity of the particle. Note that this opposing force is proportional to v.

Kirkwood and Eisenschitz used these ideas to consider the behaviour of a molecule of the liquid surrounded, now, by similar molecules of the same size. One obvious difference from the previous case of a suspended particle is that the environment of a representative liquid molecule cannot any more be regarded as being continuous. Nevertheless, Kirkwood and Eisenschitz were able to use the analogy with Brownian motion to discuss viscosity arising in steady laminar flow, and obtained a differential equation for $u(r)$; the radial distortion mentioned in Chapter 12.4. This equation is of the second order, and its solution necessitates the application of two boundary conditions. There has been some conflict of opinion as to what these should be. Suffice it to state here that a value of $u(r)$ can be obtained by solving this equation and hence a value of η calculated by using equation (12.5). The boundary conditions given by

Eisenschitz and Suddaby (1953) give a form of $u(r)$, which leads to reasonably good argeement between the value of η for liquid argon calculated from equation (12.5) and those obtained from experiment.

12.6 CONCLUSION

In this chapter we have concentrated almost entirely on the Kirkwood-Eisenschitz theory as applied to shear viscosity in liquids. The phenomenon of thermal conductivity has also been studied on the basis of this theory. Since a temperature gradient is then involved, the appropriate non-equilibrium distribution function has been obtained by assuming that the temperature varies according to

$$T = T_0 (1 + bz)$$

where the second term in the bracket takes account of the non-equilibrium distortion, the counterpart of $u(r)$ in the viscosity case. In this way Zwanzig, Kirkwood, Stripp and Oppenheim (1953) calculated the thermal conductivity of liquid argon and found it to be 2.4×10^{-4} cal cm^{-1} s^{-1} deg^{-1}, as compared with the experimental value of 2.9×10^{-4} cal cm^{-1} s^{-1} deg^{-1}.

It must be emphasized that we have given only a somewhat sketchy account of the Kirkwood-Eisenschitz theory of liquid transport processes in this Chapter, and for fuller details the reader should consult the more detailed references at the end of the Chapter. The great importance of the theory is that it is the most successful attempt to treat transport phenomena in liquids on the basis of molecular dynamics. Relatively little work has been done on transport properties of non-electrolytic solutions because the lattice model of a solution is relatively crude and the distribution-function theory of such solutions is still being developed. However, due largely to the pioneer work of Onsager, the theory of equilibrium and transport properties of electrolytes has been highly developed and very good agreement between theory and experiment is often found. For an account of this field the reader is referred to Harned and Owen *The Physical Chemistry of Electrolytic Solutions* (Reinhold, New York, 1950).

REFERENCES

Books
Cole, G. H. A., 1967, *An Introduction to the Statistical Theory of Classical Simple Dense Fluids,* Chs. 7 and 8, Pergamon Press Ltd.
Donnelly, R. J., Herman, R. and Prigogine, I. (Eds.), 1966, *Non-equilibrium Thermodynamics Variational Techniques and Stability,* University of Chicago Press.

Eisenchitz, R., 1958, *Statistical Theory of Irreversible Processes,* Oxford University Press.

Eyring, H. and Jhon, M. S., 1969, *Significant Liquid Structures,* Ch. 9, John Wiley and Sons, Inc.

Harned, R. S. and Owen, B. B., 1950, *The Physical Chemistry of Electrolytic Solutions,* Reinhold, New York.

Prigogine, I., 1955, *Introduction to Thermodynamics of Irreversible Processes,* Charles C. Thomas, Illinois, U.S.A.

Pryde, J. A., 1966, *The Liquid State,* Chs. 9. and 10, Hutchinson University Library, London.

Temperley, H. N. V., Rowlinson, J. S. and Rushbrooke, G. S., (Eds.), 1968, *Physics of Simple Liquids,* North-Holland Publishing Company, Amsterdam, Chapter 12 by P. Gray.

Papers and Articles
Brush, S. G., 1962, *Chem. Rev., 62,* 513.

Eisenschitz, R. K. and Suddaby, A., 1953, *Proc. 2nd Int. Congr. Rheol.,* p. 320, London: Butterworths Scientific Publications.

Kirkwood, J. G., 1946, *J. Chem. Phys.,* **14,** 180.

Suddaby, A., 1970, *Transport Processes in Liquids.* Article in *Liquids,* Contemporary Physics Reprint, pp. 42–60. Taylor and Francis, Ltd.

Zwanzig, R. W., Kirkwood, J. G., Stripp, K. F. and Oppenheim, I., 1953, *J. Chem. Phys.,* **21,** 2050.

Chapter 13
NON-NEWTONIAN LIQUIDS

13.1 INTRODUCTION

The study of the deformation and flow of matter is termed **rheology** and we shall devote this Chapter to a brief description of the main types of rheological behaviour.

We have already considered liquids which obey Newton's law of constant viscosity in Chapter 6. For such a Newtonian liquid the viscosity η is a constant at a given temperature and pressure and is independent of the rate of shear. The relation obeyed by such a liquid is

$$\eta = \frac{\text{shear stress } \tau}{\text{rate of shear } S}.$$

Provided the flow is laminar the graph of shear stress against the rate of shear, known as the **consistency curve** for the liquid, is a straight line whose slope is the constant viscosity η. Water is a Newtonian liquid, as are most pure single-phase liquids and solutions of substances of low molecular weights. In these liquids the viscous dissipation of energy is due to collisions between *fairly small* molecules.

There are however other types of liquids, known as non-Newtonian liquids, which do not have a constant coefficient of viscosity but one which varies with the rate of shear. Such liquids can be further divided into two groups, known as **time-independent** and **time-dependent** liquids. These two groups will now be described.

13.2 LIQUIDS OF THE TIME-INDEPENDENT TYPE

A non-Newtonian liquid of this type is characterised by the fact that its viscosity is not dependent on its duration of flow; in this respect it is like a Newtonian liquid. The consistency curves for the various types of time-independent liquids are shown in Fig. 13.1.

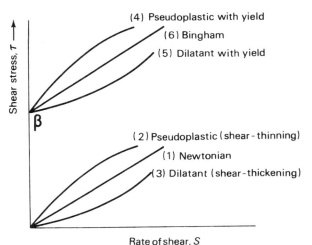

Fig. 13.1 Consistency curves for time-independent non-Newtonian liquids.

The straight line in curve 1 corresponds to a Newtonian liquid and its slope gives the value of the constant viscosity η of this liquid. In curve 2 the consistency curve for a **pseudoplastic** or **shear-thinning** liquid is shown. Since this curve has no linear portion the value of τ/S varies along it; thus there is no *constant* coefficient of viscosity for this type of liquid as there is for a Newtonian liquid. We can however take any point of coordinates (τ, S) on this curve and the quantity τ/S will give the value of the **apparent viscosity** η_a at this point. η_a will clearly vary in value as we move along the curve, and it is meaningless to quote its value unless the rate of shear corresponding to it is also stated. Liquids of this type thin upon shearing, that is the value of η_a decreases as the shear rate S increases. Examples of these shear-thinning liquids are dilute solutions of high polymers, some polymer melts like rubbers and cellulose, and suspensions such as paints and detergent slurries.

The behaviour of shear-thinning liquids is usually explained as follows. If, for example, we consider a dilute high polymer solution, it is thought that the long polymer molecules become increasingly aligned along the streamlines as the rate of shear increases. The apparent viscosity becomes less as this degree of alignment increases; when the molecular major axes eventually coincide with the streamlines the apparent viscosity thereafter attains a steady value.

Curve 3 shows the behaviour of a **dilatant** or **shear-thickening** liquid and is the opposite of curve 2. These liquids thicken upon shearing and are not as common as the shear-thinning types; their apparent viscosity increases with increasing shear rate. Collyer (1973b) quotes a good example of a shear-thickening liquid in which 66.7g of wheat starch is mixed with 100 cm^3 of water. When such a mixture is stirred slowly with a rod the viscous force on the rod is small, but if the rod is moved suddenly so as to involve a high shear rate the viscous force arrests the motion of the rod. Similarly the rod will fall easily through the liquid under gravity, but if it is suddenly plunged into the surface of the liquid its motion is arrested very quickly. There is at present no really satisfactory theory to explain the molecular mechanisms involved in shear-thickening behaviour.

For liquids described by curves 2 and 3 the relation

$$\tau = \alpha S^n$$

usually holds, where α and n are constants for the liquid concerned. For a shear-thinning liquid $n < 1$ and for a shear-thickening liquid $n > 1$; for a Newtonian liquid n is equal to unity. The apparent viscosity of a liquid described by either curve 2 or 3 is given by

$$\eta_a = \frac{\tau}{S} = \alpha S^{n-1}$$

Thus, since $n < 1$ for a shear-thinning liquid, η_a decreases as S increases. Conversely, for a shear-thickening liquid $n > 1$ and η_a increases as S increases.

There is another class of substances whose behaviour is represented by curves 4, 5 and 6 of Fig. 13.1, and these are known as materials **with a yield value**. They are a strange mixture of a solid and a liquid. Let us concentrate on the type of material represented by curve 4 which corresponds to a **pseudoplastic material with a yield value**. Such a material deforms initially like a solid until a certain value β of the shear stress is exceeded; this value β is known as the **yield stress**. For shear stresses less than β no flow will occur (that is, the material behaves like a solid) but for shear stresses exceeding this value these materials behave as normal pseudoplastic liquids. Similarly there are materials known as **dilatant with a yield value**, and these are represented by curve 5 of Fig. 13.1. Such a material also behaves as a solid for stresses less than the yield value, and as a dilatant liquid for stresses exceeding this value.

There is one interesting and special case of a material with a yield value whose behaviour is shown by the straight-line curve 6 of Fig. 13.1. It is known as a **Bingham plastic**. This material deforms elastically until the yield stress is reached, but once this stress is exceeded it flows as a Newtonian liquid with τ linearly related to S. In other words a Bingham material may be described as Newtonian with a yield value. For any point (τ, S) on the straight line 6 of slope μ we have

$$\mu = \frac{\tau - \beta}{S}$$

where μ is the Bingham viscosity for a yield value of β. The corresponding apparent viscosity η_a of the Bingham plastic is

$$\eta_a = \frac{\tau}{S} = \mu + \frac{\beta}{S}.$$

The flow of toothpaste out of its tube is a good example of Bingham plastic flow, and it will be readily appreciated that the yield value β is essential to the behaviour of the toothpaste. Other examples are drilling muds and fresh cement. Typical values of μ and β for fresh cement are 2400 mNsm^{-2} and 48 Nm^{-2} respectively.

The behaviour of Bingham plastic materials, and of close approximations to them, is usually explained as follows. For shear stresses below the yield value β such a material has a three-dimensional structure. This structure is preserved for shear stresses up to this yield value, but once the shear stress exceeds β the structure breaks down and flow starts to occur. On subsequently reducing the shear stress to values below β the structure starts to reform.

13.3 LIQUIDS OF THE TIME-DEPENDENT TYPE

A time-dependent non-Newtonian liquid is one whose viscosity depends on the duration of flow. For such liquids there is no unique relation between the shear stress and rate of shear, and so consistency curves cannot be drawn to illustrate their behaviour.

If a time-dependent liquid is made to flow under a constant rate of shear, both the shear stress and the apparent viscosity gradually decrease with time. When such a liquid is stirred or shaken it becomes 'runny' and forms a **sol**; on then being left alone it resets, forming a **gel**. Such a reversible isothermal gel-sol-gel sequence is called **thixotropy**. Perhaps the most familiar examples are thixotropic paints; after being stirred to a sol they can be readily applied to a surface where they soon reform into a gel which will not 'run'. In some cases thixotropic materials behave like a Bingham plastic in the sense that they possess a yield value below which no flow occurs.

As already stated, thixotropic liquids revert to their original gel structure on being left to stand. For some liquids however this rebuilding of their structure occurs more rapidly when the liquids are stirred gently or their containers rolled slowly in the hands. Collyer (1974) refers to these as **rheopectic** liquids. Essentially they are thixotropic liquids whose original structures are rebuilt more quickly if the process is helped along by a little gentle shearing. Such small shearing motions help to rebuild the structure once it has been destroyed, whereas larger shears destroy the structure. An interesting example of a rheopectic liquid consists of a 42 per cent gypsum paste in water. If this liquid is shaken into a sol it resets in about 40 minutes if allowed to stand undisturbed. If, instead, its container is rolled gently in the hands the liquid resets in only 20 seconds.

Much less common are **negative thixotropic** liquids which are the opposite in their behaviour to thixotropic liquids. Their apparent viscosity increases with time under a constant rate of shear and they exhibit a sol–gel–sol sequence which is reversible and isothermal.

There are several different types of thixotropic materials such as suspensions, polymer solutions and melts, and it is not possible to conceive of a single model which will explain the behaviour of them all. However we can describe a *general* kind of model which will explain the main features of thixotropic behaviour. Such a model must account for the possible existence of a yield value and also the time-dependence of the subsequent flow. Thixotropic materials are composed of large molecules usually packed fairly loosely because of their unsymmetric shapes such as thin discs, long needles and polymer chains. The presence of a yield value can be explained by assuming that there is some form of thixotropic bonding between the molecules which keeps the structure rigid up to this yield value. As soon as the flow starts, the continued decrease in apparent viscosity is attributed to the progressive alignment of the major axes of the molecules along the streamlines. This alignment requires a finite and measurable time; this is in contrast with the case of the shear-thinning liquids described in Chapter 13.2, where such alignment occurs in a very short time which cannot really be measured.

In negative thixotropic liquids it is believed that *more* intermolecular bonds are formed as a result of stirring motions as compared with the number in the undisturbed liquid. This increase in the number of bonds on shearing explains how a gel can be created.

13.4 SOME PRACTICAL APPLICATIONS OF NON-NEWTONIAN LIQUIDS

The properties of non-Newtonian liquids are of great importance in the food industry. Here are a few examples. Salad cream and ketchup are gels in their initial undisturbed state, but after shaking their containing bottles the viscosity decreases sufficiently for one to be able to pour them out. Their yield values are obviously also of importance, as is that of margarine. Chocolate and fruit gums must be firm at ordinary temperatures, but must flow in the warmer temperature in one's mouth.

Similarly toothpaste and shaving creams have to stay in their tubes but must be able to flow out when these tubes are squeezed. It is necessary that thixotropic paints in the sol state should spread easily on a surface, but they must also quickly revert to the gel state and not run over the surface. The yield value for ink in a ball-point pen is of importance since the ink must not run except when the pen is used in writing.

In oil drilling work the fluid used to lubricate the drill is called a drilling mud. In drilling a well this mud is pumped down inside the drill along its axis, and it is then forced upward in the annular gap between the drill and the cylindrical wall of the well. The main function of the mud is to lift drilling debris from the drill head to the top of the well. The yield value is of great importance here: if the drill is stopped the drilling debris will tend to fall down the well unless the mud sets at once as in a Bingham type mud, or at least fairly quickly, as in a thixotropic mud of the right consistency. What is really required is that the debris remains in suspension and does not settle down at the bottom of the well.

For further details of these applications of non-Newtonian liquids the reader is referred to the articles by Collyer listed at the end of the Chapter.

13.5 VISCOELASTIC LIQUIDS

There is another class of liquids which possess elastic as well as viscous properties, and they are known as **viscoelastic** liquids. Such a material is mainly liquid-like in its behaviour, yet has some of the characteristics of an elastic solid. For example, if it is sheared it is able to store elastic energy just as a solid does.

Many solutions and melts of high polymers with very long molecules (macromolecules) are viscoelastic. Their viscous behaviour is related to the degree of alignment of the macromolecules with the streamlines and depends on the bonds between these molecules. Their elastic behaviour depends on the internal structure of the macromolecules. Each of these has a number of 'rigid' links and each link can take up various orientations with respect to its neighbours. This enables each molecule to assume a large number of shapes as the material is deformed elastically.

13.6 SOME EFFECTS SHOWN BY VISCOELASTIC LIQUIDS

We shall now consider briefly some of the best-known effects exhibited by viscoelastic liquids, most of which are shown very well by solutions of poly-ethylene oxide. These solutions are easily prepared and in what follows we shall be mainly concerned with their behaviour in particular.

The **elastic recovery** of a viscoelastic liquid is easily observed by pouring the liquid from one beaker into another. If the pouring is stopped for a moment and the vertical column of liquid is snipped sharply with a scissors, the upper part of the column 'recovers' quickly by recoiling suddenly into the upper beaker. The so-called **Weissenberg effect** occurs when a vertical cylindrical rod is immersed in a viscoelastic liquid and is then rotated rapidly. It is found that the liquid climbs up the rotating rod for quite a distance. Collyer has recently shown this Weissenberg effect even more convincingly by having a layer of viscoelastic liquid 'sandwiched' between two Newtonian liquids. When a glass rod is rotated in this three-layered arrangment the viscoelastic liquid climbs up and down the rod into both Newtonian liquids. The usual explanation of this effect is as follows. The streamlines in the viscoelastic liquid around the rotating rod are horizontal concentric circles whose centres lie on the axis of rotation of the rod. It is also known (see for example Trevena, 1968) that, in a viscoelastic liquid, the stresses along the streamlines and those normal to the streamlines, in this instance the vertical direction, are not equal. This difference between the stresses is known as the **normal-stress difference**, and it causes the viscoelastic liquid to climb up and down the rotating rod.

Another effect caused by these normal-stress differences is the **die-swell** effect, which occurs when a viscoelastic liquid flows downwards through a tube of circular cross-section. In this case stresses are set up normal to the streamlines, which are parallel to the axis of the tube. When the liquid emerges at the lower end of the tube these normal stresses are relieved by a considerable increase in diameter of the emerging stream. The increase of diameter can be two or three times the diameter of the tube and is called a **die-swell**. This effect is important in the manufacture of man-made fibres and plastics.

Kaye (1963) found that if a certain solution of polyisobutylene in dekalin were poured into a dish then, when the falling stream was thin enough, the stream would bounce off the liquid in the dish every few seconds and rise in an arc before falling back into the dish. This **Kaye effect** has been re-examined recently by Collyer and Fisher (1976).

13.7   THEORETICAL WORK ON VISCOELASTIC LIQUIDS

The theoretical treatment of the behaviour of viscoelastic liquids, which are materials possessing both elasticity and viscosity, involves the use of a *complex* shear modulus G^*. To illustrate this let us consider a viscoelastic liquid being subjected to a small-amplitude sinusoidal stress in a coaxial cylinder viscometer. This consists of a hollow outer cylinder and a suspended inner cylinder with the liquid occupying the gap between them. If the outer cylinder is rotated sinusoidally through a small angle $\phi = \phi_0 \sin \omega t$, the resulting movement of the liquid will cause the inner cylinder to oscillate with the same angular frequency ω. There will, however, be a phase difference δ between the oscillations of the two cylinders, so that the equation $\theta = \theta_0 \sin(\omega t + \delta)$ represents the movement of the inner cylinder. If the space between the cylinders were occupied by a perfectly elastic solid, the two cylinders would oscillate in phase, that is $\delta = 0$. In the other extreme case, if this space were filled by a Newtonian (that is, purely viscous) liquid, then it is found that $\delta = \pi/2$ (see, for example, Trevena, 1975).

To return to the first case in which we have a viscoelastic liquid in the viscometer, the phase angle δ ($0 < \delta < \pi/2$) is a measure of the elasticity of the liquid. In the course of each swing, a part of the energy is stored elastically with the increase of amplitude and is recovered on the counter-swing as this amplitude decreases. The rest of the energy is lost as heat due to the viscous forces, and this is not recovered. The 'real' elastic modulus G' is a function of the energy stored in the swing, and the 'viscous' modulus G'' a function of the energy lost as heat. Together they are combined to give a complex modulus $G^* = G' + iG''$ and the phase angle $\delta = \tan^{-1}(G''/G')$. For a purely elastic solid $G'' = 0$ and $\delta = 0$; for a purely viscous liquid $G' = 0$ and $\delta = \pi/2$. Since, dimensionally, [shear modulus] = [viscosity/time], we can write $G' = \eta'\omega$ etc., and the result $\eta^* = \eta' + i\eta''$ also holds where η^* is the **complex dynamic viscosity**.

The above experiment involves one of the most simple types of motion of a viscoelastic liquid. To study the behaviour of such a liquid when subjected to more complicated types of motion it is necessary to formulate equations of state for the liquid. Such equations have to satisfy two principles, one involving memory and the other certain invariance conditions. The principle involving memory is that the behaviour of an element of the liquid depends only on its previous deformation history and not on the state of neighbouring elements. The principle concerned with invariance conditions demands that the behaviour of an element of the material does not depend on the motion of the liquid as a whole through space. In other words, this principle when thus stated refers to an indifference to the motion of the liquid as a whole. It can also be restated on the basis of a principle of indifference to an observer identified with a moving co-ordinate frame, although Walters (1965) has shown that both approaches are equivalent.

These two principles were used by Oldroyd to consider the invariant forms of equations of state for viscoelastic liquids which were applicable under all conditions of motion. In his approach he used a convected co-ordinate system drawn in the liquid and deforming continuously with it. He also used the second principle in the form independent of the motion of the liquid as a whole through space. Other approaches based on indifference to an observer have been discussed by Rivlin and Ericksen, Green and Rivlin, and Coleman and Noll (see Trevena, 1968, for further details).

In conclusion, one can state that the theoretical understanding of viscoelastic liquids is now in a fairly well-advanced state and that this knowledge should help the experimentalist and technologist concerned with these liquids to make further advances in the next few years.

REFERENCES

Books
Lodge, A. S., 1964, *Elastic Liquids,* Academic Press.
Scott Blair, G. W., 1969, *Elementary Rheology,* Academic Press Inc. Ltd., London.
Trevena, D. H., 1975, *The Liquid Phase,* The Wykeham Science Series.
Walters, K., 1975, *Rheometry,* Chapman and Hall.

Papers
Collyer, A. A., 1973a, *Physics Education,* **8**, 111.
Collyer, A. A., 1973b, *Physics Education,* **8**, 333.
Collyer, A. A., 1974a, *Physics Education,* **9**, 38.
Collyer, A. A., 1974b, *Physics Education,* **9**, 313.
Collyer, A. A. and Fisher, P. J., 1976, *Nature,* **261**, 682.
Kaye, A., 1963, *Nature,* **197**, 1001.
Trevena, D. H., 1968, *Contemp. Phys.* **9**, 537.
Walters, K., 1965, *Nature,* **207**, 826.

Chapter 14
LIQUID HELIUM

14.1 INTRODUCTION

It is well known that liquid helium behaves very differently from other liquids, and we shall devote this Chapter to summarizing its main properties. First however we shall summarize some of the properties of the element helium itself. Naturally occurring helium has an atomic weight of just less than four and consists of two isotopes. Each atom of the abundant isotope has a nucleus consisting of two protons and two neutrons, whilst each atom of the rare isotope has a nucleus made up of two protons and one neutron. They are written respectively as He^4 and He^3. To every one atom of He^3 in the earth's atmosphere there are about ten million atoms of He^4, so that He^3 is very rare. Each helium atom of either kind has two electrons which, in the lowest energy state, occupy the first shell around the nucleus.

At ordinary temperatures helium is a monatomic gas. After various unsuccessful attempts by different workers, Kamerlingh Onnes successfully liquefied helium in 1908. In appearance liquid helium is colourless and clear, and under atmospheric pressure it boils at 4.2K. Helium has a critical temperature of 5.2K, and as it is cooled further it undergoes a transformation at 2.17K. This temperature of 2.17K is called the λ-point, for reasons which will be given in Chapter 14.3. It is found that below the λ-point liquid helium displays many properties not found in other liquids. The liquid phases above and below the λ-point are known respectively as He I and He II. When we say 'liquid helium II' it should be emphasized that we are, in fact, referring to the heavy isotope He^4 only. This is because there was no experimental evidence for any transition in the light isotope He^3 until recently. For this reason we shall discuss the properties of liquid helium (He^4) first, and then discuss the light isotope (He^3) briefly in Chapter 14.10.

14.2   THE (PRESSURE, TEMPERATURE) CHARACTERISTICS

The equilibrium (pressure, temperature) characteristics of helium are shown in Fig. 14.1. The diagram shows clearly the existence of four phases. These are

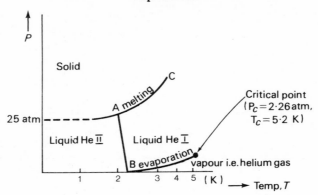

Fig. 14.1 The phase diagram for helium 4.

the solid phase, two liquid phases (consisting of two phases He I and He II), and the gaseous, or vapour, phase. The presence of a critical point on the liquid-vapour curve is shown as in a normal liquid. So are the melting and evaporation curves which define respectively the solid–liquid and liquid–vapour boundaries. However there is no triple point in the normal sense, that is there is no point where the melting and evaporation curves intersect. In other words there is no temperature and pressure at which the solid, liquid and gaseous phases can exist together in equilibrium. There are however two 'triple' points lying at A, B, the ends of the λ-line separating the two liquid phases. At the point A solid helium, He I and He II exist in equilibrium; at B, He I, He II and helium gas exist in equilibrium.

Helium will remain liquid under its own vapour pressure down to 0K. In order to solidify helium it is necessary to apply an external pressure. This can be seen if we follow the solid-liquid line from around the point C in Fig. 14.1 as the temperature decreases. At about 1.8K the melting curve rapidly flattens and approaches nearly horizontally the value of 25 atm at absolute zero. This implies that an external pressure of 25 atm is necessary to solidify liquid helium at 0K. In fact helium is the only substance which cannot be solidified under its own vapour pressure simply by reducing the temperature. These facts are explained as being due to the high zero-point energy of helium (see Chapter 14.4).

The thermodynamics of the process shows that as the temperature becomes less than that corresponding to the point A in Fig. 14.1, where the λ-line meets the melting curve, the internal energy of liquid helium is less than that of solid helium (see, for example, F. London, *Superfluids,* Vol. II, Chapter 1). This is why the liquid is the stable phase under these conditions. Helium is unique in this respect. From the Clapeyron equation (see equation (5.2)) we have

$$\frac{dP}{dT} = \frac{S_{\text{liquid}} - S_{\text{solid}}}{V_{\text{liquid}} - V_{\text{solid}}}$$

$$= \frac{\Delta S}{\Delta V} \tag{14.1}$$

where dP/dT is the slope of the melting curve and ΔS, ΔV are the differences in entropy and volume respectively of the two phases before and after melting. As $T \rightarrow 0$, $dP/dT \rightarrow 0$; ΔV is found to remain finite and has a constant value at the low temperatures concerned. Hence ΔS tends to zero, in agreement with the Third Law of Thermodynamics.

The melting of helium at absolute zero, indeed up to 1K, is different from that of all ordinary substances. Since ΔS is zero there is no latent heat. If the liquid is compressed above $P = 25$ atm it is isothermally solidified; if the pressure is removed it liquefies. Since there is no latent heat the melting process is a *purely mechanical* one and the relation

$$\Delta U = -P \Delta V \tag{14.2}$$

holds were ΔU is the internal energy difference between liquid and solid helium.

14.3   THE LAMBDA TRANSITION

As liquid helium is cooled below 2.17K it starts to expand instead of continuing to contract. This fact was first observed by Kamerlingh Onnes and Boks in 1924 (see Fig. 14.2). At the λ-point the density reaches a maximum and there is also a discontinuity in the slope of the (density, temperature) curve, implying a discontinuity in the coefficient of thermal expansion at the λ-point as well. This coefficient becomes negative below 2.17K and becomes positive again below 1.15K.

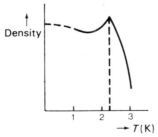

Fig. 14.2 The density of liquid helium 4 as a function of temperature

A sharp anomaly in the specific heat at the λ-point was found by Keesom and Clusius in 1932 (see Fig. 14.3). There is a very rapid rise in the curve up to the λ-point and then a very rapid fall. The specific heat then decreases slowly until about 2.5K afterwards rising more gradually. The transition temperature is called the λ-point because the curve resembles the letter 'λ'. The curve shows the variation with temperature of the specific heat of liquid helium in equilibrium with its own vapour. If the pressure is increased the λ-point temperature is lowered. This variation of the λ-point with pressure is shown as the λ-line **AB** in Fig. 14.1.

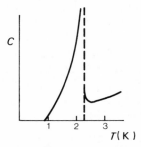

Fig. 14.3 The Lambda Transition showing the temperature variation of the specific heat of liquid helium 4 under its own vapour pressure.

In addition to the specific heat anomaly the λ-transition is accompanied by anomalies in the expansion coefficient (already mentioned) and the isothermal compressibility, κ_T, of the liquid. The transition is not accompanied by a latent heat.

14.4   THE STRUCTURE OF LIQUID HELIUM NEAR 0K: ZERO POINT ENERGY

Liquid helium is a 'simple' liquid in the sense that it consists of monatomic molecules which are spherically symmetrical. In this case the mutual potential energy $\phi(r)$ of any two molecules is a function of their separation r only, and the variation of $\phi(r)$ with r follows the general shape shown in Fig. 3.1b.

Various attempts have been made to obtain the form of $\phi(r)$. One function which has been widely used is the *LJD* function of equation (3.31), namely

$$\phi(r) \; = \; 4\epsilon \left[\left(\frac{\sigma}{r} \right)^{12} - \left(\frac{\sigma}{r} \right)^{6} \right]$$

where $-\epsilon$ is the minimum value of $\phi(r)$ at $r = r_0$ and σ is the molecular diameter. The values of ϵ and σ for helium are

$$\epsilon \; = \; 14.11 \times 10^{-23} \text{ J}$$

$$\sigma \; = \; 2.556 \times 10^{-10} \text{m}.$$

These values have been obtained by comparing the experimental value of the second virial coefficient with that obtained from the *LJD* function. Better values of the function $\phi(r)$ have been given by Yntema and Schneider, Slater and Kirkwood and by de Boer and Michels. For a fuller discussion of these functions the reader is referred to *Superfluids*, Volume II by F. London. All these functions have the general shape shown in Fig. 3.1b.

To discuss the structure of liquid helium below the λ-point it is best to start at the absolute zero of temperature and consider *solid* helium first. It is known from X-ray analysis that solid helium forms a hexagonal close-packed lattice. Knowing $\phi(r)$ it is then possible to calculate the potential energy per mole of the solid; this lattice energy turns out to be −62 calories per mole at 0K. On classical grounds this would be the only energy we would have to consider because each atom would possess zero kinetic energy at 0K. According to quantum theory however, each atom is regarded as vibrating in three independent directions and even at 0K each of these three modes of vibration has a minimum kinetic energy of $\frac{1}{2}h\nu$, where ν is the frequency of oscillation and h is Planck's constant. This minimum energy is called the **zero point energy** (see Chapter 3.3). The zero point energy for solid helium is +50 cal/mole (see, for example, Atkins' *Liquid Helium,* Ch. 2). Thus the resultant internal energy of solid helium is −62 +50 = −12 cal/mole. We therefore see that at 0K the zero point energy is numerically about 80 per cent of the lattice energy and tends to oppose the effect of the attractive forces.

Let us now, following F. London, apply similar ideas to *liquid* helium. This involves estimating the lattice potential energy and the zero point energy. Let us consider each in turn.

(a) The lattice potential energy per mole, Φ

London found this by assuming both a definite lattice structure and a form for $\phi(r)$. He was thus able to obtain a curve showing the variation of Φ with molar volume V at 0K and this curve is the dotted curve (1) in Fig. 14.4.

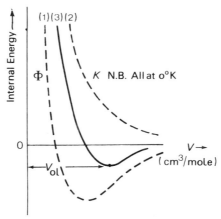

Fig. 14.4 F. London's treatment of the energy of liquid helium.

(b) The zero point energy per mole, K

Estimating this is more difficult, but it can be done by considering each atom as being in a spherical cell formed by its nearest neighbours. The behaviour of the enclosed atom is then considered on the basis of the Schroedinger equation and a value for its zero point energy in the ground state is obtained. From this the total zero point energy per mole, K, is found. The variation of K with V at 0K is shown in curve 2 of Fig. 14.4.

The total internal energy is then given by curve 3, the resultant of curves 1 and 2. The equilibrium volume $V_{o\varrho}$ of liquid helium at zero pressure and temperature is that corresponding to the minimum of curve 3 and is almost three times the volume corresponding to the minimum of curve 1. So the zero point energy can be regarded as equivalent to an extra repulsive force between the atoms of the liquid, which is thereby 'inflated' to a larger volume.

Fig. 14.5 shows, for comparison, the total internal energies of solid and liquid helium at 0K plotted against the molar volume.

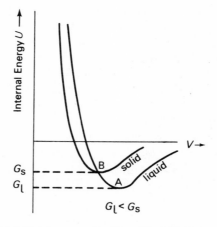

Fig. 14.5 Total energy of solid and liquid helium.

The Gibbs free energy G is given by

$$G = U - TS + PV$$

and a particular phase is in equilibrium when G is a minimum. When $T = 0$, as is the case for Fig. 14.5, the condition for equilibrium is that $(U + PV)$ should be a minimum. However, if the solid and liquid are also under the saturated vapour pressure only, then $P \to 0$ as $T \to 0$, and the minimum of U gives the equilibrium condition. This is the case at the minima A and B of the two curves in Fig. 14.5 where $P = -\partial U/\partial V = 0$ and the two corresponding tangents are horizontal, cutting the U-axis as shown. The intercepts they make on this axis are G_ϱ and G_s the values of G for the liquid and solid respectively, and since $G_\varrho < G_s$ the liquid is the stable phase.

Consider now the case of an externally applied pressure. Since $P = -\partial U/\partial V$, the equation of the tangent to either curve in Fig. 14.5 is of the form

$$U + PV = U'$$

where U' is the intercept of the tangent with the U-axis. As P increases from zero $\partial U/\partial V$ decreases and the tangents to both curves swing around in a clockwise direction. At first the intercepts $U_\varrho' = G_\varrho$ and $U_s' = G_s$ with the U-axis are such that $G_\varrho < G_s$ until we reach the condition shown in Fig. 14.6. Here the two curves have a common tangent and in this case $G_\varrho = G_s$, both liquid and solid phases coexisting in equilibrium. If the pressure is raised still further

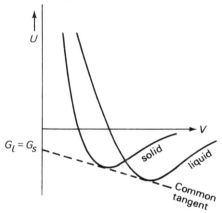

Fig. 14.6 The case in which the solid and liquid curves have a common tangent.

G_ϱ becomes greater than G_s and the solid becomes the stable phase. The above argument shows then how the application of pressure produces solidification. Without such an external pressure the stable phase at the absolute zero is the liquid not the solid. This occurs because the zero-point energy inflates the volume so much that the liquid, rather than the solid, is the stable phase. Indeed, this is the essence of London's theory.

Zero point energy is really a consequence of the Uncertainty Principle, as may be seen from the following argument. The helium atoms are pictured as occupying a lattice of sites and each normal mode of the lattice behaves as a three-dimensional harmonic oscillator. The mean energy of a one-dimensional oscillator is, from quantum mechanics, given by

$$E = \tfrac{1}{2}h\nu + \frac{h\nu}{e^{h\nu/kT} - 1} \tag{14.3}$$

where ν is the frequency of the oscillator (see, for example, Chapter IV of *The Third Law of Thermodynamics* by J. Wilks). The first term on the right hand side is independent of the temperature and is the zero point energy while the second term depends on T and is the thermal energy. As $T \to 0$, the zero point energy does not change but the thermal energy tends to zero. So as the temperature decreases the loss of thermal energy means that the spatial positions of the atoms become more defined. Hence by the uncertainty principle the uncertainty in the momentum of the atom increases. Even at absolute zero, when the atoms have lost all their thermal energy, each normal mode is still excited to an extent represented by the zero point energy $\frac{1}{2}h\nu$. Since at these very low temperatures the quantum mechanical zero point energy is such an important factor in the behaviour of liquid helium, the latter is often referred to as a **quantum liquid**.

14.5 SUPERFLUIDITY AND THE TWO-FLUID MODEL

There are two well known methods of measuring the viscosity η of a liquid. The first involves measuring the damping of the torsional oscillations of a disk suspended in the liquid by a fibre, while the second involves measuring the velocity of flow of the liquid through narrow pipes or gaps under a certain pressure head (see Chapter 6.8.2). Let us now consider the results obtained from trying to apply these methods to obtain the viscosity of liquid helium II.

The oscillating disk method gives a value of the order of 10^{-5} poise for the viscosity of the liquid. When the second method is employed however, a new feature of liquid helium II known as **superfluidity** is observed. This superfluidity is manifested by a much lower viscosity and was reported simultaneously in two adjacent letters in *Nature* by Kapitza (1938) and by Allen and Misener (1938). In Kapitza's experiments the viscosity was measured by the pressure difference accompanying the flow of the liquid between two optically flat glass surfaces separated by about 0.5×10^{-4} cm. In their experiments, Allen and Misener observed the flow of the liquid through long capillaries under a pressure head. The values of the viscosity reported by these workers were of the order of 10^{-11} poise, and Allen and Misener further reported that the flow velocity seemed almost independent of the pressure head. For comparison we should remember that η is about 10^{-2} poise for water, 2×10^{-5} poise for liquid helium I, and 2×10^{-4} poise for air.

The difference in the viscosities of liquid helium II obtained by the two methods can be explained on the basis of the two-fluid model due to Tisza (1940) and Landau (1941). According to this model, liquid helium II is regarded as a mixture of two interpenetrating fluids known as the normal and superfluid components. The normal component has the viscosity of helium I and other properties expected of a normal liquid with high zero point energy at low temperatures. It is the viscosity of this normal component which is measured by the oscillating disk method. The superfluid component is very different: it has virtually no viscosity and no entropy and can flow through very narrow tubes and gaps at very high velocities. The two fluids do not interact with each other and the superfluid does not interact with the walls of the vessel, capillary or gap in contact with it. Each fluid in fact behaves as though the other were not present.

The presence of these two components was verified in an experiment due to Andronikashvili (1946) in which a pile of disks, separated by gaps of about 0.2 mm, were made to perform torsional oscillations in liquid helium (Fig. 14.7a). Above the λ-point all the liquid in the gaps between the disks moved with the disks and therefore contributed to the moment of inertia of the rotating system. Below the λ-point on the other hand, only the normal component was dragged around by the disks because the superfluid did not interact frictionally with the disks. As the temperature continued to fall below the λ-point the frequency of the torsional oscillations increased, indicating that a greater fraction of the liquid was behaving as a superfluid and a smaller fraction was carried round by the disks. Finally at absolute zero it *all* behaved as a superfluid. The results of this experiment are shown in the curves of Fig. 14.7b, where the fractions ρ_n/ρ and ρ_s/ρ of the normal and superfluid components respectively are plotted versus temperature below the λ-point. At any particular temperature the density ρ of helium II is the sum of the densities of the two components; thus $\rho = \rho_n + \rho_s$.

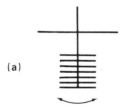

(a)

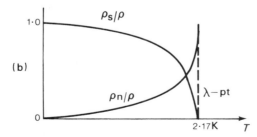

(b)

Fig. 14.7 Andronikashvili's experiment.

The two-fluid model must not be regarded as a complete theory of liquid helium II, but it does describe the superfluid properties discussed in the present Section. Historically, this superfluidity was first observed by Keesom and van den Ende (1930). They noticed how liquid helium II passed through tiny leaks which neither liquid helium I nor gaseous helium would pass through above the λ-point. More detailed theories of liquid helium are discussed in Chapter 14.11.

14.6   THE LIQUID HELIUM FILM

When a solid is partially immersed in liquid helium II the exposed part of its surface soon becomes covered by a film of the liquid. Among the early experiments which illustrate the properties of such films are those due to Daunt and Mendelssohn (1939a). If an empty beaker is partially immersed in a bath of the liquid, there is a flow of the liquid via the surface film until the liquid levels inside and outside the beaker are the same (Fig. 14.8a). If this beaker is then raised, the direction of flow in the film is reversed until the levels inside and outside are again equal (Fig. 14.8b). If the beaker is raised clear of the surface in the bath it empties itself as the film flows down the outside surface of the beaker forming drops of liquid which then fall into the bath (Fig. 14.8c). A closer investigation of these processes shows that the volume transfer of the liquid per

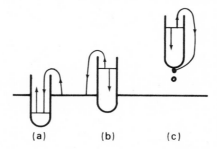

| (a) | (b) | (c) |

Fig. 14.8 Film flow experiments of Daunt and Mendelssohn.

unit width of the film occurs at a constant rate for a given temperature. This transfer rate is also only weakly dependent on the length of the distance travelled, on the difference between the liquid levels, and on the height of the barrier over which the liquid climbs.

The fact that the rate of flow is virtually independent of the pressure head and the length of the film suggests the possible existence of a terminal critical flow velocity v_c. The observations suggest that the film starts to flow with an initial acceleration up to this velocity v_c and that thereafter it moves with this critical velocity. This is further supported by the fact that the volume rate of transfer is proportional to the smallest perimeter of the connecting surface above *both* free surfaces of the liquid. This was demonstrated by Daunt and Mendelssohn using the apparatus shown in Fig. 14.9. Provided the outside level of the liquid was higher than that of the constriction in the inner beaker, the volume rate of transfer R_1 between the vessels had a certain constant value. As soon as this outside level fell below the constriction, this rate fell to a lower value R_2 such that

$$\frac{R_1}{R_2} = \frac{\text{inside circumference of inner beaker}}{\text{perimeter of constriction}}$$

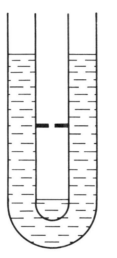

Fig. 14.9 The effect of a constriction on film flow.

The volume rate of transfer, which is very dependent on temperature, rises from zero at the λ-point to about $R = 7.6 \times 10^{-5}$ cm^3 per sec per cm of surface width at 1.5K, corresponding to a value of ν_c of about 30 cm per sec. Below this temperature R is fairly constant, but below 0.8K it increases again.

The above behaviour of liquid helium II can be explained on the basis of its superfluid properties. Let us refer again to Fig. 14.8a. The parts of the beaker surface which are initially exposed are in equilibrium with the saturated vapour of the liquid. This leads to the formation of an *adsorbed* layer or film of helium on this surface whose thickness is about 2×10^{-6} cm. The mobile adsorbed film stretches continuously between the inside and outside liquid levels and acts as a 'conveyor belt' enabling the two levels to become equal. The liquid in the film flows very much as if it were a superfluid with complete absence of friction.

14.7   THERMAL EFFECTS

Consider helium II at a particular temperature T: if heat is supplied, the temperature rises above T and the concentration ρ_n/ρ of the normal fluid increases while that of the superfluid decreases, and vice-versa.

Suppose that we start with two vessels A and B, joined by a fine capillary, in which the temperature and liquid levels are initially the same (Fig. 14.10). We have seen previously that the superfluid component can move freely through such a capillary whereas the normal component cannot; thus the capillary behaves rather like a semi-permeable membrane. If heat is now supplied to the liquid in B, the concentration of the normal fluid in B rises and an osmotic pressure is produced with the result that some of the superfluid tends to flow

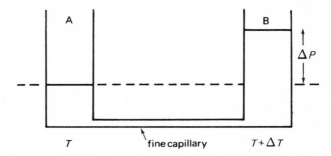

Fig. 14.10 The thermomechanical effect.

from A to B equalizing the concentrations in both vessels. However, as this flow from A to B occurs it causes a hydrostatic pressure difference ΔP to build up, and the flow ceases when ΔP is equal and opposite to the osmotic pressure. Let us emphasize in particular that the flow of liquid is towards the supply of heat.

This movement of liquid and build-up of pressure is known as the **thermomechanical** or **fountain effect**. The magnitude of the effect was derived by H. London (1939) by assuming that the transfer of the superfluid between the two vessels does not involve any entropy change and is therefore a thermodynamically reversible process. London's result is

$$\frac{\Delta P}{\Delta T} = \frac{S}{V} = \rho S \qquad (14.4)$$

where S and V are the entropy and volume of unit mass of the bulk liquid. We can also derive (14.4) by regarding ΔP as the **osmotic pressure** proper to the change in 'concentration' ρ_n/ρ produced by the temperature difference ΔT.

Experiments to verify (14.4) have been made by Kapitza (1941) using apparatus essentially like that shown in Fig. 14.11. The vessel A of Fig. 14.10 is replaced by a bath of helium, vessel B by the inside of a dewar flask, and the connecting capillary by the gap between two optical flats F_1, F_2. This gap is sufficiently small to inhibit the passage of the normal fluid through it. The

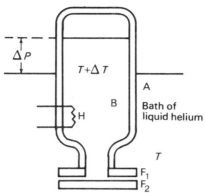

Fig. 14.11 Kapitza's apparatus for measuring the thermomechanical effect.

coil shown supplies the heat to the liquid inside the flask. The results show that
the relation (14.4) holds to within about 2 per cent.

A striking illustration of the thermomechanical effect was observed by
Allen and Jones (1938) using the apparatus of Fig. 14.12. The lower bent end of
a glass tube is packed tightly with emery powder and acts as a semi-permeable
membrane. When a beam of light is shone on to the system, the powder absorbs
heat, the temperature inside the tube rises, and superfluid rushes in violently so
as to emerge as a fountain or jet as shown. Jets as high as 30 cm have been
observed.

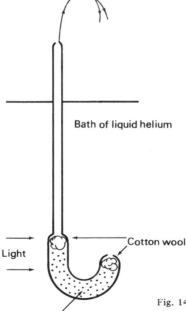

Fig. 14.12 The fountain effect
(after Allen and Jones).

Since the thermomechanical effect is thermodynamically reversible, a converse mechanocaloric effect in He II must be possible also. This was discovered by Daunt and Mendelssohn (1939b) using the apparatus of Fig. 14.13. A is a dewar vessel with an exit O at its lower end blocked with emery powder. As the liquid flows downwards through this powder plug the temperature of the liquid remaining in A rises. Since the superfluid is the only component which flows out, the superfluid concentration in A falls. This in turn causes some of the normal fluid in A to change into superfluid so as to preserve the appropriate concentration. This change from normal fluid to superfluid results in a liberation of heat which causes the temperature in A to rise.

The thermal effects described above are due to the very effective mode of heat transfer in helium II. Before discussing this however, let us compare the thermal conductivities of helium I and helium II. The thermal conductivity of the former has a value similar to that of an ordinary liquid; its value is 4.5×10^{-5} cal/cm s K at the λ-point rising to 6.3×10^{-5} cal/cm s K at the boiling point. The apparent thermal conductivity of helium II can however reach very much higher values, sometimes of the order of a thousand times that of copper. This large change in conductivity below the λ-point is the reason

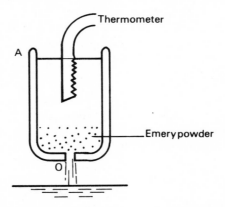

Fig. 14.13 Apparatus to demonstrate mechanocaloric effect (after Daunt and Mendelssohn).

for the marked difference in the visual appearance of the liquid below this temperature. Above the λ-point helium I boils briskly with the formation of bubbles; below this temperature it is quiet and still. This is because the heat transfer is so rapid that bubbles have no time to form and all the evaporation occurs at the free surface. The term 'thermal conductivity' of helium II must be used with reserve because experiment shows that the heat transfer depends on the temperature gradient and the geometry of the apparatus in ways unlike all other liquids. In fact, it is really better to talk about thermal *convection*.

To explain the above thermal effects we can invoke the two-fluid model.

In helium II the normal fluid and superfluid are completely interpenetrating. When heat is supplied the superfluid moves to the warmer regions and becomes normal fluid. Conversely, the normal fluid will move away from the warmer regions if it can, though this may be impossible if it has to flow through a narrow capillary or tightly packed powder. Thus one of the most important factors which account for the extremely high apparent thermal conductivity of helium II is the high mobility of the superfluid enabling it to circulate very rapidly through the bulk liquid.

14.8   FIRST AND SECOND SOUND

We would expect that an ordinary sound wave could be propagated through liquid helium II. However, because of the two-fluid composition of the liquid a second type of wave propagation, known as **second sound** is also possible.

Ordinary or 'first' sound is a density wave because variations of density are propagated by it. In this case the normal and superfluid components move together in phase. At any point the total density ρ varies, whilst there is only a small, second order change in the relative concentrations of the two components at that point. The velocity c_1 of this first sound is given, just as for an ordinary fluid, by

$$c_1{}^2 = (\partial P/\partial \rho)_S .$$

In second sound the normal and superfluid components at a point move in opposite directions in such a way that, while there is only a small change in the density, a larger first order change occurs in the relative concentrations, ρ_n/ρ_s, of the two components. Now the normal component is hot and the superfluid cold, so at points where the concentration of the normal fluid increases we have a higher temperature and, conversely, a lower temperature implies a higher superfluid concentration. This can be illustrated as in the sine curve of Fig. 14.14 where temperature at any particular instant is plotted against distance. At the crests A, B, C etc. the normal component has its maximum concentration and the temperature is a maximum; at the troughs D, E, F, etc. the superfluid concentration is a maximum and the temperature a minimum. Thus second sound is a temperature wave or, in other words, propagation of temperature differences at fixed density ρ. The velocity of second sound is given by

$$c_2{}^2 = \frac{\rho_s}{\rho_n} S^2 \left(\frac{\partial T}{\partial S}\right)_\rho \tag{14.5}$$

where S refers to unit mass of the bulk liquid. This result follows if the two fluids move independently of each other.

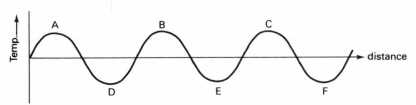

Fig. 14.14 Temperature wave occurring in second sound.

To sum up: first sound is a wave of density fluctuations under virtually adiabatic conditions and second sound is a wave of entropy fluctuations under conditions of virtually constant density. Density and pressure in the former play roles similar respectively to entropy and temperature in the latter.

Second sound was first predicted by Tisza (1940) and its existence verified and its velocity measured by Peshkov (1944) using a 'transmitter' and 'receiver' placed at a certain distance apart in the liquid. The transmitter was a flat spiral of wire which fed periodic heat waves into the liquid. The receiver was another spiral through which a steady d.c. current initially passed. When the second sound wave reached the receiver a voltage oscillation was set up which could be observed on an oscilloscope. By moving the receiver a standing wave distribution of temperature could be set up, nodes and antinodes being formed. Thus the velocity c_2 and the wavelength could be measured. c_2 turns out to be of the order of 20 m/s in the temperature range 1.4 to 1.8K. The corresponding value of c_1 is about 240 m/s, about ten times higher. This shows clearly that second sound propagation is quite distinct from that of first sound. It is also different from the propagation of periodic temperature differences in an ordinary conducting medium. No matter how high the conductivity, there is always considerable damping of the wave over one wavelength for ordinary conduction.

The apparently strange idea that something as chemically simple as helium can behave like a mixture of two distinct fluids is correct. We can measure ρ_s/ρ_n either *mechanically*, as in Andronikashvili's experiment (Chapter 14.5), or *thermally* from the velocity of second sound using equation (14.5). The two measurements agree very closely, although it is hard to think of two more dissimilar ways of measuring the same physical quantity. As Landau, F. London, and others have shown, a satisfactory interpretation of the physical meaning of ρ_s/ρ_n is obtained from quantum mechanics, giving us further confidence that it is a real physical quantity. The idea that one fluid has high mobility and negligible entropy is also in accord with quantum mechanics.

14.9   CAVITATION IN LIQUID HELIUM II

It is only comparatively recently that cavitation in liquid helium II has been studied, and values for the critical tension needed to produce cavitation have been given by Misener and Hebert (1956) and Beams (1956, 1959). These values, obtained by essentially static methods, are summarized in Table 14.1. More recent experimental work by Finch *et al* (1964, 1966) and Finch and Wang (1966) has also been reported.

Table 14.1.

Experimenter	Method	Critical tension (atm)
Misener and Hebert (1956)	Bellows	0.30
Beams (1956)	Spinning capillary	0.14
Beams (1956)	Piston and cylinder	0.07

There seems to have been very little work done to estimate the *theoretical* critical tension of helium II. However Geraghty and Trevena (1969) calculated the energy required to produce an empty cavity in liquid helium II as a function of the radius of the cavity. To do this they used the experimentally determined pair potential energy function $\phi(r)$ for helium together with information about the structure of the liquid as derived from the measured radial distribution function.

14.10 LIQUID HELIUM 3

Helium 3 is the light isotope whose nucleus contains one neutron and two protons. The critical temperature and normal boiling point for liquid helium 3 are respectively $T_c = 3.32K$ and $T_B = 3.19K$. The liquids helium 3 and 4 show similar behaviour near their respective critical temperatures, the behaviour of both liquids in the critical region being very like that of ordinary liquids. When helium 3 originally became available for experiments one looked for properties like the remarkable ones shown by helium 4, but it was soon found that it remained apparently a normal liquid down to temperatures well below the lambda point of helium 4. Liquid helium 3 has a finite viscosity which increases with decreasing temperature very much like an ordinary liquid. Since its nucleus has a finite magnetic moment its magnetism can be studied by resonance methods, but it appears to behave like an ordinary paramagnetic body. The fact that helium 3 was behaving quite normally at temperatures well below the lambda point of helium 4 was accepted as confirming that the remarkable properties of helium 4 were of quantum-mechanical origin. It was also correctly predicted that because the helium 3 nuclei obey Fermi-Dirac statistics while helium 4 nuclei obey Bose-Einstein statistics, the transition of helium 3 and the behaviour of the low temperature phases might be entirely different. Because of the high zero-point energy, helium 3 also remains liquid down to absolute zero unless pressures of the order of 30 atmospheres are applied.

No transition was found down to 0.01K so it was realised that the transition, if it existed at all, would need refined methods of magnetic cooling to detect it. Various claims to have observed a transition were made, but this was definitely detected by Osheroff *et al* (1972), the transition temperature being at about 2 millidegrees K. See also J. C. Wheatley (1973). It has been found that there is not one superfluid phase but two, and, since the nuclei have a magnetic moment, the magnetic field can be used as a thermodynamic variable as well as pressure and temperature. The variety of phenomena discovered in the last few years has been quite bewildering, and we can do no more than refer the reader to a review by Leggett (1976).

A further wide variety of phenomena is presented by solutions of helium 3 in helium 4, study of which has helped to elucidate the properties of both liquids and particuarly those of helium 4.

Helium 3 occurs naturally in the atmosphere and also in natural gas from wells, but always in very small concentration compared with helium 4. By far the most important source is the decay of radioactive tritium prepared in atomic reactors.

14.11 THEORIES OF HELIUM 3 AND 4

The two-fluid model of liquid helium II is helpful in correlating many of the unusual properties of this liquid. It is clearly correct in principle, but it remains intellectually unsatisfying unless we can suggest a physical reason for the existence of the two fluids. Actually, the two successful theories are based on treatments very similar to those introduced in Chapter 2.2 and 2.3, where the liquid is treated as a dense gas or as a loosely bound solid.

In the gas-like approach, we begin with an ideal gas of He^4 nuclei. These should obey Bose-Einstein statistics since they each contain four nucleons. Neglecting for the moment their mutual interactions, the quantum states that can be occupied by such particles confined in a cubical box of side d are found from Schroedinger's equation to be

$$\epsilon = \frac{h^2}{8md^2} (p^2 + q^2 + r^2) \tag{14.6}$$

where h is Planck's constant, m the mass of the helium atoms and p, q, r are integers (not zero). According to Bose-Einstein statistics, the expected number of particles in a quantum state of energy ϵ_i is

$$n_i = \frac{1}{\lambda\, e^{\epsilon_i/kT} - 1} \tag{14.7}$$

where λ has to be chosen so that n_i, summed over all permissible quantum states, is equal to the total number of atoms in the assembly. It was pointed out by Einstein as a theoretical curiosity that, below a certain temperature depending on the density, such a gas would have a finite fraction of all its atoms in the very lowest possible quantum state, thus $p = q = r = 1$ in (14.6). If the temperature dropped still lower, the fraction of atoms in the very lowest state would increase. If we introduced more atoms into the assembly at a constant volume and temperature they would *all* go into the very lowest state, leaving the occupation of the others unchanged. This strange result is a mathematical consequence of the expressions (14.6) and (14.7) together with the physical requirement that all the n_i must be positive. For a fuller discussion the reader is referred to works on statistical mechanics. See, for example J. Wilks' *The Third Law of Thermodynamics*, Ch. 6.

The condensation temperature is given by

$$T_c = \frac{h^2}{2\pi m k} \left[\frac{N}{2.612V} \right]^{2/3}. \tag{14.8}$$

If we insert for N/V the value appropriate to the observed density of liquid helium we calculate $T_c = 3.14\text{K}$ compared with the observed transition temperature of liquid He II of 2.17K.

This fairly close agreement led F. London to suggest in 1938 that the transition of liquid helium was essentially of this type but was modified by the fact that real helium atoms interact strongly with one another at such densities. Indeed there would be no liquid if they did not! One can go on to argue that the atoms in the lowest quantum state might be expected to have very unusual properties, such as making zero contribution to the entropy of the assembly. It can also easily be shown that this lowest quantum state is drastically modified by the effect of gravity, so that we can understand the flow of the superfluid under very small pressure gradients. Adsorbed films of helium would be expected to be very mobile and sensitive to gravity as well.

In the alternative, solid-like, approach Landau used a picture of the liquid very similar to that described in Chapter 7.8, where the motions of the molecules are described as equivalent to a superposition of sound waves of various frequencies, just as they are in a solid. He investigated the effect of the transverse motion of a boundary wall of the containing vessel on such an assembly, and concluded that at very low temperatures there would be some 'slip' in the sense that only a portion of the mass of such an assembly would be carried along by the wall. Again this leads to a two-fluid picture. Two types of excitation are postulated in the liquid helium II. The first is a **phonon**, quite similar to the sound quanta associated with the acoustic frequencies that can exist in the liquid. The liquid can possess other modes of motion, each involving only a few helium atoms, and the corresponding excitation was known as a **roton**. Evidence for the existence of both these types of excitation has been obtained by neutron scattering experiments by Palevsky *et al* (1957) and by Yarnell *et al* (1959). The final picture is that of a ground-state liquid with zero-point energy in which these excitations can be thermally excited, the excitations comprising the normal fluid. If the liquid is heated more excitations come into existence, thus increasing the concentration of the normal fluid. Landau's theory does not predict the lambda point, and can be regarded as complementing F. London's theory.

Another approach, more intellectually satisfying than either of the above, is to work without a preconceived model at all, but to try to solve the Schroedinger equation for N interacting helium atoms directly. This leads to the very difficult many-body problem in quantum mechanics, but Feynman (1953, 1954, 1955) was able to by-pass many of the difficulties. He showed that the many-body wave-equation can be transformed in such a way as to relate the energy-spectrum of possible excitations to the liquid distribution function. This (see Chapter 10) can be regarded as an observable quantity which can be fed into the theory instead of using the interaction function. He obtained an energy spectrum agreeing quite reasonably with experiment.

Much work on liquid helium 4 is still being carried out and interesting facts are still being discovered, particularly boundary effects and effects associated with critical velocities. However there does not seem to be any real reason to doubt that they will all eventually be shown to be logical consequences of Schroedinger's many-body equation, and that no further postulate is needed.

Can we say the same of liquid helium 3? At the time of writing it is much too early to make such a synthesis. When the superfluid properties were discovered in a Fermi type assembly it was natural to look to the Bardeen-Cooper-Schrieffer theory of superconductivity for guidance. The details are beyond the scope of this book, but the fundamental concept can be described fairly easily. This is the 'Cooper pair' of electrons being correlated in momentum but not in position. Although this concept is foreign to classical mechanics, it occurs naturally as a consequence of quantum mechanics. It is already clear that this concept throws much light on the behaviour of superfluid helium 3, but whether other fundamental innovations are needed also remains to be seen. The reader is referred to A. J. Leggett's review article (1976).

REFERENCES

Books

Atkins, K. R. *Liquid Helium,* Cambridge University Press, 1959.

Jackson, L. C., *Low Temperature Physics,* New York, John Wiley and Sons, Inc., 1954.

Keesom, W. H., *Helium,* Amsterdam, Elsevier Publishing Co., 1942.

London, F., *Superfluids,* Vol. II, New York, John Wiley and Sons, Inc., 1954.

Mendelssohn, K., *Cryophysics,* New York, Interscience Publications, Inc., 1960.

Mendelssohn, K., *The Quest for Absolute Zero,* World University Library, 1966.

McClintock, M., *Cryogenics,* New York, Reinhold Publishing Corporation, 1964.

Simon, F. E., Kurti, N., Allen, J. F. and Mendelssohn, K., *Low Temperature Physics,* London, Pergamon Press, Inc., 1952.

Temperley, H. N. V., *Changes of State,* Cleaver-Hume Press, 1956.

Wilks, J., *The Third Law of Thermodynamics,* Oxford University Press, 1961.

Wilks, J., *The Properties of Liquid and Solid Helium,* Oxford, Clarendon Press, 1967.

Papers

Allen, J. F., Peierls, R. and Uddin, M. Z., 1937, *Nature,* **140,** 62.

Allen, J. F. and Misener, A. D., 1938, *Nature,* **141,** 75.

Allen, J. F. and Jones, H., 1938, *Nature,* **141,** 243.

Adronikashvili, E. L., 1946, *J. Phys. Moscow,* **10,** 201.

Beams, J. W., 1956, *Phys. Rev.* **104,** 880.

Beams, J. W., 1959, *Phys. Fluids,* **2,** 1.

Betts, D. S., 1969, *Contemp. Phys.,* **10,** 241.

Daunt, J. G. and Mendelssohn, K., 1939a, *Proc. Roy. Soc.,* **A170,** 423, 439.

Daunt, J. G. and Mendelssohn, K., 1939b, *Nature,* **143,** 719.

Feynman, R. P., 1953, *Phys. Rev.,* **91,** 1301.

Feynman, R. P., 1954, *Phys. Rev.,* **94,** 262.

Feynman, R. P., 1955, *Prog. Low Temp. Phys.,* **1,** 17, editor Gorter (North Holland).

Finch, R. D., Kagiwada, R., Barmatz, M. and Rudnick, I., 1964, *Phys. Rev.,* **134A,** 1425.

Finch, R. D. and Wang, T. G. J., 1966, *J. Acoust. Soc. Am.,* **39,** 511.

Geraghty, D. and Trevena, D. H., 1969, *J. Phys. C.,* **2,** 1877.

Kapitza, P. L., 1938, *Nature,* **141,** 74.

Landau, L. D., 1941, *J. Phys. Moscow,* **5,** 71.

Keesom, W. H. and Clusius, K., 1932, *Proc. Roy. Acad. Amsterdam,* **35,** 307.

Leggett, A. J., 1975, *Rev. Mod. Phys.,* **47,** 331.

Leggett, A. J., 1976, *Endeavour,* **35,** 83.

London, H., 1939, *Proc. Roy. Soc.*, **A171**, 484.

Misener, A. D. and Hebert, G. R., 1956, *Nature*, **177**, 946.

Osheroff, D. D., Gully, W. J., Richardson, R. C. and Lee, D. M., 1972, *Phys. Rev Lett.*, **28**, 885; **29**, 920.

Palevsky, H., Otnes, K. and Larsson, K. E., 1958, *Phys. Rev.* **112**, 11.

Peshkov, V. P., 1944, *J. Phys. Moscow*, **8**, 131, 381.

Tisza, L., 1940, *J. Phys. Radium*, 1, **165**, 350.

Wheatley, T. C., 1973, *Physica*, **69**, 218.

Wheatley, T. C., 1976, *Rev. Mod. Phys.*, **47**, 415.

Appendix 1
THE RULE OF EQUAL AREAS
AND THE VAN DER WAALS' ISOTHERM

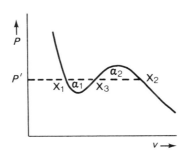

Fig. A1

Referring to Fig. A1, choose two points X_1, X_2 as shown such that $X_1 X_2$ is parallel to the v-axis. If the pressure P' at which the change of phase from liquid to vapour in equilibrium with it is to be the value of P at X_1 and X_2, then the position of the line $X_1 X_2$ is determined from three facts:

(1) At equilibrium the Gibbs functions, g, per unit mass of liquid and vapour must be equal, by the general results of thermodynamics.

(2) The thermodynamic relation $(\partial g/\partial P)_T = v$ holds, where v refers to unit mass. (See equations 13 in Appendix 2).

(3) The same relation between P and v holds for the liquid and vapour branches of the curve in Fig. A1.

Thus if g_1, g_2 denote the values of g at X_1, X_2 respectively, we can write

$$g_2 - g_1 = \int_{X_1}^{X_2} v dP \qquad (A1.1)$$

$$= \int_{X_1}^{X_3} v dP + \int_{X_3}^{X_2} v dP \qquad (A1.2)$$

$$= \alpha_1 - \alpha_2$$

(using the appropriate sign convention)

where α_1 and α_2 are the areas shown. The fact that $g_1 = g_2$ at equilibrium implies that $\alpha_1 = \alpha_2$, which is the Maxwell 'rule of equal areas'. Note that it only holds if condition (3) above is satisfied.

SUMMARY OF FORMULAE IN
THERMODYNAMICS AND STATISTICAL MECHANICS

An assembly consisting of N particles (that is atoms or molecules) occupying a volume V can be treated from the point of view of thermodynamics or statistical mechanics. There are various equations which form a 'bridge' between the two approaches. Perhaps the most celebrated is

$$S = k \ln W, \tag{A2.1}$$

which links the entropy S of the assembly and the number of complexions W of the assembly. Equations (A2.17) and (A2.20) below are further examples of such equations.

First, consider an assembly consisting of several components $1, 2, \ldots, i$ etc. such that there are N_1 particles of type 1, etc., that is, $\sum_i N_i = N$. The assembly has certain **extensive properties** which are proportional to the size of the assembly, that is to the N_i's. Such properties include the volume V, entropy S and internal energy U. Other properties, known as **intensive properties**, are independent of the size of the assembly. Examples of these are the external pressure P, temperature T and chemical potentials μ_i.

Suppose first that the numbers N_i are fixed and that a quantity of heat dQ is absorbed by the assembly. Then by the first law of thermodynamics we can write

$$dQ = dU + PdV. \tag{A2.2}$$

Here dU is the change in internal energy and $-PdV$ is the work done by the assembly against the external pressure. Further, if the change is reversible then by the second law of thermodynamics

$$dQ = TdS. \tag{A2.3}$$

So from (A2.2) and (A2.3) we have

$$dU = TdS - PdV \tag{A2.4}$$

for this small reversible change of state.

In addition, if each of the numbers N_i changes to $N_i + dN_i$, we must also add a term $\mu_i dN_i$ proportional to dN_i for each i to account for such changes. If we do this we have

$$dU = TdS - PdV + \sum_i \mu_i dN_i \qquad (A2.5)$$

All we say about the μ_i's at this stage is that they are some intensive properties of the assembly; we shall show later that they are the chemical potentials.

Suppose now that we picture our final assembly as being built up in stages from nothing in such a way that the intensive properties T, P and μ_i on the r.h.s. of (A2.5) remain constant. Then we need to sum the various equations of type (A2.5) over all these stages to obtain our equation representing the final state of the assembly, so we write $\Sigma dU = U$, $\Sigma dS = S$, $\Sigma dV = V$ and $\Sigma dN_i = N_i$. Thus we obtain

$$U = TS - PV + \sum_i \mu_i N_i . \qquad (A2.6)$$

The Helmholtz free energy A and the Gibbs free energy G are defined as

$$A = U - TS \qquad (A2.7)$$

and

$$G = A + PV. \qquad (A\,2.8)$$

So from (A2.6), (A2.7) and (A2.8) we get

$$G = \sum_i \mu_i N_i . \qquad (A2.9)$$

This shows that the μ_i's are the partial Gibbs free energies or chemical potentials.

If we consider simultaneous small changes in all the quantities in equation (A2.7) we have

$$dA = dU - TdS - SdT$$

which, with (A2.5) gives us

$$dA = -SdT - PdV + \sum_i \mu_i dN_i. \qquad (A2.10)$$

Hence

$$\left(\frac{\partial A}{\partial T}\right)_{V,N_i} = -S, \left(\frac{\partial A}{\partial V}\right)_{T,N_i} = -P, \left(\frac{\partial A}{\partial N_i}\right)_{T,V} = \mu_i. \qquad (A2.11)$$

Again, considering small changes throughout equation (A2.8) gives us

$$dG = dA + PdV + VdP$$

which with (A2.10) gives us

$$dG = -SdT + VdP + \sum_i \mu_i dN_i. \qquad (A2.12)$$

Hence

$$\left(\frac{\partial G}{\partial T}\right)_{P,N_i} = -S, \left(\frac{\partial G}{\partial P}\right)_{T,N_i} = V, \left(\frac{\partial G}{\partial N_i}\right)_{T,P} = \mu_i. \qquad (A2.13)$$

Consider next an assembly consisting of N *identical* particles each of mass m in a volume V. If the assembly is a perfect gas there will be no interactions between the individual particles, and the partition function Z for the whole assembly is

$$Z = \left(\frac{2\pi mkT}{h^2}\right)^{3N/2} \frac{V^N}{N!} \qquad (A2.14)$$

where $f = \left(\frac{2\pi mkT}{h^2}\right)^{3/2} V$ is the partition function for the translational motion of a single particle of the assembly.

When the interaction forces between the particles are also taken into account the partition function Z in (A2.14) becomes modified to

$$Z = \left(\frac{2\pi mkT}{h^2}\right)^{3N/2} Q(N) \qquad (A2.15)$$

where $Q(N)$ is the configurational partition function which depends on the 'configuration', that is on the coordinates of the particles. It is given by

$$Q(N) = \frac{1}{N!} \underset{\leftarrow 3N \rightarrow}{\int \ldots \int} e^{-\Phi/kT} dx_1 \ldots dz_N. \qquad (A2.16)$$

In this equation Φ is the total potential energy of the assembly given by

$$\Phi = \sum_{i<j} \phi(r_{ij})$$

where $\phi(r_{ij})$ is the potential energy of the pair ij of particles.

The Helmholtz free energy of the assembly is then given by

$$A = -kT \ln Z \tag{A2.17}$$

and its internal energy by

$$U = A + TS$$

$$= A - T\left(\frac{\partial A}{\partial T}\right)_V$$

$$= -T^2 \frac{\partial}{\partial T}\left(\frac{A}{T}\right)_V \tag{A2.18}$$

Equation (A2.18) is the Gibbs-Helmholtz formula. From (A2.17) and (A2.18) we get

$$U = kT^2 \left(\frac{\partial \ln Z}{\partial T}\right)_V \tag{A2.19}$$

$$= \frac{kT^2}{Z}\left(\frac{\partial Z}{\partial T}\right)_V . \tag{A2.20}$$

REFERENCES

Rushbrooke, G. S., 1949, *Statistical Mechanics,* Appendix I, Oxford.
Mayer, J. E. and M. G., 1948, *Statistical Mechanics,* Appendix VIII, John Wiley.

Index